KU-517-944

6·50

THE THYML

IN FOCUS®

Titles published in the series:

*Antigen-presenting Cells

*Complement

DNA Replication

Enzyme Kinetics

Gene Structure and Transcription

Genetic Engineering

*Immune Recognition

*B Lymphocytes

*Lymphokines

Membrane Structure and Function

Molecular Basis of Inherited Disease

Protein Biosynthesis

Protein Engineering

Protein Targeting and Secretion

Regulation of Enzyme Activity

*The Thymus

*Published in association with the British Society for Immunology.

Series editors

David Rickwood

Department of Biology, University of Essex, Wivenhoe Park,
Colchester, Essex CO4 3SQ, UK

David Male

Institute of Psychiatry, De Crespigny Park, Denmark Hill,
London SE5 8AF, UK

THE THYMUS

Mary A.Ritter

Department of Immunology, Royal Postgraduate Medical School,
Hammersmith Hospital, Du Cane Road, London W12 0NN, UK

I.N.Crispe

Immunobiology Section, Yale Medical School,
310 Cedar Street, New Haven, CT 06510, USA

OXFORD UNIVERSITY PRESS

1659714

Oxford University Press
Walton Street, Oxford OX2 6DP

Oxford is a trade mark of Oxford University Press

Published in the United States
by Oxford University Press, New York

© Oxford University Press 1992

All rights reserved. No part of this publication may be reproduced, stored in a
retrieval system, or transmitted, in any form or by any means, electronic,
mechanical, photocopying, recording, or otherwise, without the prior permission
of Oxford University Press.

This book is sold subject to the condition that it shall not, by way of trade or
otherwise, be lent, re-sold, hired out, or otherwise circulated without the
publisher's prior consent in any form of binding or cover other than that in
which it is published and without a similar condition including this condition
being imposed on the subsequent purchaser.

A catalogue record for this book is available from the British Library

Library of Congress Cataloging in Publication Data
Ritter, Mary A.
The thymus / Mary A. Ritter, I.N. Crispe.
(In Focus)
Includes index.
1. Thymus.   I. Crispe, I. N.   II. Title.   III. Series: In focus
(Oxford, England)
[DNLM: 1. Thymus Gland.   WK 400 R615t]
QR185.8T48R58   1992   599'.142—dc20   91-7108
ISBN 0-19-963144-1 (pbk.)

BIRMINGHAM UNIVERSITY LIBRARY

Typeset and printed by Information Press Ltd, Oxford, England.

Preface

There is never an ideal time to review a fast-moving subject, but the present would seem to be particularly appropriate for writing a book on the thymus, since the topic can now be viewed from the new vantage-point of knowledge that accrued from the explosive burst of thymic research during the latter half of the 80s. Such a quantum jump in the understanding of intrathymic events, especially those controlling repertoire selection, was made possible by a combination of imaginative ideas and new technical approaches, prominent amongst the latter being the development of monoclonal antibodies to individual T cell receptor $V\beta$ gene products and the construction of elegant transgenic mice that expressed defined components of either an antigen-specific T cell receptor or the MHC restriction element.

However, despite the many advances in our understanding of T cell development within the thymus, there are clearly areas that are still poorly understood. For example, how do thymocytes interact with their microenvironment and what is the nature of the microenvironmental signals that regulate gene expression in T cells at successive development stages as they pass through their characteristic intrathymic locations? We hope that these and other outstanding problems will be solved for the next edition of this book.

Finally, we would like to thank our many colleagues with whom we have discussed the content of this volume. In particular, we thank those who either read some of our early efforts or generously provided photomicrographs to illustrate some of the chapters (Andrew Farr, Dimitris Kioussis, Heather Ladyman, Marion Kendall, Willem van Ewijk, Brita von Gaudecker, Anne Wilson).

Mary A.Ritter
I.N.Crispe

Contents

4. The T cell repertoire 41

5. The thymic microenvironment 57

6. Unresolved issues in thymus research

Abbreviations

APC	antigen presenting cell
B cell	bone marrow derived lymphocyte (mammals)
	bursa of Fabricius-derived lymphocyte (birds)
CD	cluster of differentiation
CGRP	calcitonin gene related peptide
Con A	concanavalin A
CTES	cluster of thymic epithelial staining
CTL	cytotoxic T lymphocyte
DC	dendritic cell
DN	double negative, CD4$^-$,CD8$^-$
DP	double positive, CD4$^+$,CD8$^+$
DTH	delayed type hypersensitivity
FITC	fluorescein isothiocyanate
GM-CSF	granulocyte-macrophage colony stimulating factor
G-v-HR	graft-versus-host reaction
H-2A	mouse MHC class II
H-2D	mouse MHC class I
H-2E	mouse MHC class II
H-2K	mouse MHC class I
HEV	high endothelial venule
HLA	human leucocyte antigen (human MHC)
HSA	heat stable antigen
I-A	mouse class II MHC; now renamed H-2A
I-E	mouse class II MHC; now renamed H-2E
IDC	interdigitating cell
Ig	immunoglobulin
IL	interleukin
IL-2R	interleukin 2 receptor
IL-4R	interleukin 4 receptor
Ifn	interferon
LAK	lymphokine activated killer cell
M-CSF	macrophage colony stimulating factor
MHC	major histocompatibility complex
MIF	macrophage migration inhibition factor
MLR	mixed lymphocyte reaction
NP-Y	neuroactive peptide-Y

pCTL	precursor cytotoxic T lymphocyte
PCV	post-capillary venule
PHA	phytohaemagglutinin
PNA	peanut agglutinin
SRBC	sheep red blood cells
T_C	T cytotoxic cell
T cell	thymus-derived lymphocyte
TCR	T cell receptor
TdT	terminal deoxynucleotidyl transferase
T_H	T helper cell
TNC	thymic nurse cell
TNF	tumour necrosis factor
TP-5	thymopentin-5
T_S	T suppressor cell
VIP	vasoactive intestinal peptide
V, D, J genes	variable, diversity and joining region genes of the TCR and Ig loci
Zn	zinc

1

Introduction

The thymus plays a central role in the creation of a fully functional immune system. Its major function is to provide the appropriate milieu within which cells of the T lymphocyte lineage can develop, proliferate, mature, generate their antigen receptor repertoire, and, at a population level, become MHC restricted and tolerant to self (*Figure 1.1*). Such cells then migrate to the periphery where they form part of the recirculating pool of lymphocytes that guard the body from invading pathogens. All vertebrate species possess a thymus. Although this may vary in anatomical position its basic cellular structure is very similar in all species studied; a conservation of structure that reflects importance of function.

In the past, the thymus has had many functions attributed to it, including that of secreting the elixir of life/eternal youth. Its relationship to the production of lymphocytes was first recognized in the eighteenth century by Thomas Hewson and his colleague Magnus Falconer (1). They proposed that the thymus was a gland that secreted fluid containing 'numberless small solid particles, similar to those found in the lymphatic glands' (lymph nodes), and that 'the lymphatic vessels arising from the thymus convey this secreted fluid through the thoracic duct into the blood vessels'. Over 100 years elapsed before this hypothesis was extended by Beard (2) when he proposed that the thymus was the parent source of leucocytes for all other lymphoid structures and that these leucocytes did 'useful work for themselves and for the body'.

However, subsequent experimental studies generated conflicting data. For example, Fagraeus (3) demonstrated that the thymus could not produce antibodies and did not respond to antigenic stimulation, while Hammar (4) and Harris and colleagues (5) found that removal of the thymus (thymectomy) from adult mice had no effect on antibody production. Thus the thymus appeared to play no direct role in an immune response. In contrast, Good and colleagues (6) observed that the thymus underwent rapid involution in response to acute infection and that immunodeficient patients with acquired agammaglobulinaemia frequently developed thymomas—indicating that the thymus was linked to the lymphoid system and immune response.

1

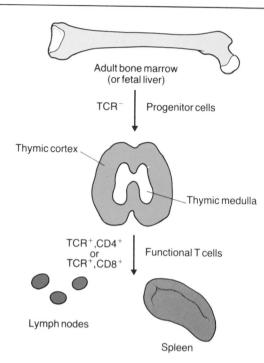

Figure 1.1.   Input and output of the thymus. The thymus is seeded with progenitor cells which come from the bone marrow in the adult, and from the liver during fetal development. Small numbers of progenitor cells enter the thymus without antigen-specific T receptors (TCR); larger numbers of functional T cells leave.

It was not until 1961 that the immunological function of the thymus was recognized. Jacques Miller, working on a murine leukaemia that originated in the thymus, showed that thymectomy of newborn mice resulted in animals that were immunodeficient. They could not reject skin allografts and could not produce effective antibodies to either sheep red blood cells (SRBC) or to salmonella H antigens (their choice of T dependent antigens was serendipitous); immunological competance could be restored by a syngeneic thymus graft (7). Similar studies confirmed these findings and showed that the delayed-type hypersensitivity response to tuberculin was also impaired (8,9). At the same time, Robert Good, in clinical studies, observed that thymic abnormalities were associated with immune disorders in man (10). These classical studies laid the foundation for modern cellular immunology.

It was subsequently shown that in birds not only did thymectomy have a similar effect, but that extirpation of a second lymphoid organ—the bursa of Fabricius— also led to immune deficits (11). However, in contrast to thymectomy, bursectomy ablated all antibody responses but did not affect graft rejection. It was then appreciated that there were two major lymphoid organs responsible for the production of lymphocytes, and the terms 'T' and 'B' lymphocyte were coined for cells produced in the thymus and bursa respectively. Although no strict bursal equivalent exists in mammals (fetal liver and adult bone marrow subserve the same function) it was recognized that a similar dichotomy of lymphocytes existed, with B cells being responsible for humoral (antibody) mediated immunity and T cells responsible for cell-mediated (graft rejection) immunity. Analysis of peripheral lymphoid organs (such as lymph nodes and spleen) in thymectomized and bursectomized animals showed characteristic areas of lymphoid depletion—giving rise to the concept of distinct T and B cell zones within these organs. Elegant cell labelling studies revealed that mature lymphocytes form a recirculating pool in the periphery (12). Cells enter the blood stream via the thoracic duct, pass through the periarteriolar sheaths of the spleen, enter the lymphatics at the lymph nodes (after passage from high endothelial venules), pass through the lymphatics to the thoracic duct and so return to the blood. During this process, lymphocytes leave the recirculating pool at the different peripheral lymphoid organs along their route and remain there for several hours before rejoining the circulation. This optimizes the opportunity for antigen and antigen-specific lymphocytes to encounter each other.

These data gave rise to the concept of primary and secondary lymphoid organs, with the primary organs (thymus and bursa of Fabricius) being responsible for the production and export of lymphocytes, while the secondary organs provided sites for the generation of immune responses to extrinsic antigen. Primary organs can be characterized by the following features: they are the first sites of lymphopoiesis in the embryo; they have a high mitotic rate independent of antigenic stimulation; and their extirpation leads to immune deficits and characteristic lymphopaenia in the secondary lymphoid organs.

The identity and source of the thymocyte progenitor remained unclear until the 1960s, with proponents of both haemopoietic and epithelial origins (2,13,14). The matter was resolved in experiments using a combination of cell markers, *in vitro* thymic organ culture and *in vivo* grafting techniques (15,16). Stem cells, originating in haemopoietic sites such as the avian yolk sac, mammalian fetal liver, and adult bone marrow, were shown to migrate via the blood stream to the thymus where, under the inductive influence of the thymic stromal microenvironment, they underwent proliferation and matured into characteristic small lymphocytes. A comparison of the effects of vascular thymic grafts and thymic tissue enclosed in diffusion chambers showed that the process requires both direct and physical contact with the thymic stroma as well as the production of soluble thymic factors (17).

Although the effects of thymectomy were found to be most marked in the suppression of cell-mediated immunity (graft rejection), there was also a reduction in the antibody response to the antigens tested, as discussed above. It was subsequently realized that T and B cells do not function in isolation, and that although B lymphocytes are the cells that produce antibody, an enhanced secondary response requires the co-operation of T lymphocytes—T cell help. Thus, heavily irradiated mice given both syngeneic bone marrow and thymus cells gave a greater antibody response to SRBC than could be accounted for by summating the effects of each population alone (18). The nature of this synergy was studied in further experiments in which heavily irradiated and thymectomized hosts were given either bone marrow or thymocytes/thoracic duct cells (a source of T lymphocytes) alone or in combination. Although it was the bone marrow cells that produced antibodies to the SRBC antigen, this response was enhanced ten-fold when thoracic duct cells were also given (19).

The following two decades saw the development of key new techniques and scientific discoveries. The use of first alloantisera and later monoclonal antibodies in immunostaining techniques revealed a wide array of cell-surface molecules such as Thy-1, CD4 (mouse, L3T4; rat W3/25), CD8 (mouse, Ly-2,3; rat, Ox8), and many others (to be discussed in subsequent chapters) (20; for details of latest workshop on human system see ref. 21). The combination of these techniques with functional studies was used to demonstrate that T lymphocytes develop into different subpopulations that can be distinguished by their phenotype, showing ultimately that the majority of CD4$^+$ T cells have helper function while the majority of CD8$^+$ T lymphocytes are cytotoxic cells (22,23).

The quest for the elusive T cell antigen receptor ended in the early 1980s when a combination of monoclonal antibody, T cell cloning and gene cloning techniques demonstrated that, although quite distinct from B lymphocyte immunoglobulin, it was clearly closely related in both structure and in the genetic events that lead to the creation of the diverse T cell repertoire (24,25,26). This creation of T cell antigen receptor diversity by gene rearrangement occurs within the thymic microenvironment.

Functional studies in genetically inbred strains of mice revealed that T lymphocytes can only respond to antigen (virus in the original experiments) when present on cells carrying the same MHC molecules; reconstitution experiments showed that this 'MHC restriction' was learnt by T cells as they matured within the thymus (*Figure 1.2*) (27). Subsequently, it has been shown that CD4$^+$ T lymphocytes see antigen associated with MHC class II while CD8$^+$ T lymphocytes see antigen associated with MHC class I molecules and that this phenomenon also arises as part of intrathymic education (*Figure 1.3*). In addition to this 'positive' selection of self MHC restricted lymphocytes within the thymus, a second, 'negative', phase of selection occurs intrathymically. First proposed by Burnet (28) as the 'clonal selection' theory, and recently demonstrated in elegant studies using monoclonal antibodies to individual T cell receptor β chains and transgenic mice (29,30,31), autoreactive T cells are deleted as they develop within the thymus. Peripheral tolerance is probably maintained by different mechanisms such as anergy and suppression.

Thus, those T lymphocytes that emerge from the thymus are functionally mature, MHC restricted, and self tolerant. They have also undergone differentiation into the two main peripheral T cell subpopulations, expressing either CD4 or CD8. Mature T cells exert both regulatory and effector functions in immune responses. CD4$^+$ cells include the T helper cells (T$_H$) which secrete growth and maturation factors (cytokines) required by B cells and other T cells to enable them to mount effective immune responses—the basis of cell co-operation. Cytokines released by CD4$^+$ cells can also attract and activate antigen non-specific accessory cells such as mononuclear phagocytes, thereby facilitating the destruction of intracellular pathogens. A minor population of

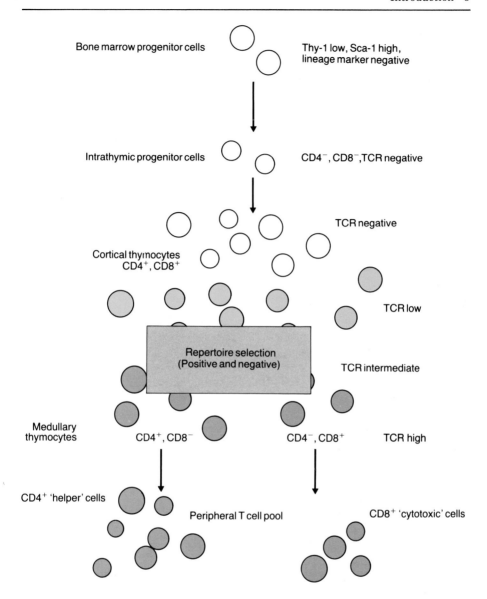

Figure 1.2. T cell lineage relationships. Within the thymus, progenitor cells which lack antigen-specific receptors (open circles) proliferate, express CD4 and CD8, and then express antigen receptors at low density (fine shading). After selection on the basis of receptor specificity, the cells up-regulate antigen receptor expression (heavy shading), and leave the thymus.

CD4$^+$ cells can act as cytotoxic cells (T$_C$), destroying target cells expressing foreign peptide antigen associated with MHC class II molecules. CD8$^+$ T cells are the principal cytotoxic T cell population, whose primary function is to kill

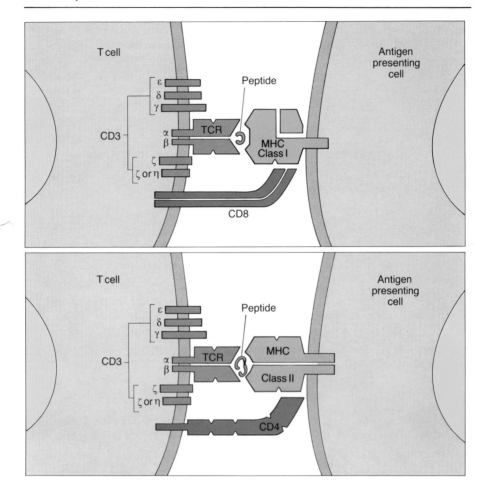

Figure 1.3.   T cell recognition of antigen involves an interaction between the variable α and β chains of the T cell receptor (TCR), and an antigenic peptide presented in a groove in the membrane-distal surface of a major histocompatability complex (MHC) molecule. Recognition also involves the co-receptor molecules CD4 and CD8, which interact with MHC class II or class I respectively. CD4, CD8, and the CD3 complex of transmembrane proteins are thought to be involved in signalling.

virus infected host cells (where viral peptides are associated with MHC class I molecules). Finally there are proposed to be CD8[+] T suppressor cells, but their identity is still a matter of hot dispute (31,32).

T lymphocytes therefore play a key role in the control and mediation of almost all immune responses (with the exception of the antibody response to T independent antigens). The thymus, as the primary source of these T cells, is therefore of paramount importance; a point emphasized by the fact that congenital absence of this organ leads to life-threatening immune deficits. Although thymic function is probably most crucial during fetal and early post-natal life when the immune system is being established *de novo*, it is likely that it is important throughout life since, even though reduced in size, the thymus in old age still contains all the essential elements identified at earlier stages. Once the immune system has been established, the small adult organ probably has sufficient capacity to generate new lymphocytes to 'top up' and maintain the repertoire of the recirculating peripheral T cell pool.

General texts

Alberts,B., Bray,D., Lewis,J., Raff,M., Roberts,K., and Watson,J.D. (1983) *Molecular Biology of the Cell*. Garland, New York.
Dixon,F.J. (ed.) (1987) *Adv. Immunol.*, **41**.
Klein,J. (1990) *Immunology*. Blackwell Scientific Publications, Boston.
Male,D., Champion,B., Cooke,A., and Owen,M.J. (1991) *Advanced Immunology*. Gower, London. 2nd edn.
Paul,W.E. (ed.) (1989) *Fundamental Immunology*. Raven Press, New York.
Silverstein,A.M. (ed.) (1989) *A History of Immunology*. Academic Press, New York.

Further reading

Cantor,H. and Weissman,I.L. (1976) Development and function of subpopulations of thymocytes and T lymphocytes. *Prog. Allergy,* **20**, 1.
Kendall,M.D. and Ritter,M.A. (ed.) (1988, 1989, 1990, 1991) *Thymus Update* (Annual Review Series), Harwood Academic, London.
Muller-Hermelink,H.K. (ed.) (1986) The human thymus. *Curr. Topics Pathol.* **75**.
Ritter,M.A. and Larche,M. (1988) T and B cell ontogeny and phylogeny. *Curr. Opin. Immunol.,* **1**, 203.

References

1. Falconer,M. (1777) *Experimental Enquiries: Part 3. Containing a description of the red particles of the blood in the human subject and in other animals; with an account of the structure and offices of the lymphatic glands, of the thymus gland, and of the spleen.* T. Longman, London.
2. Beard,J. (1900) *Anat. Anz.,* **18**, 515.
3. Fagraeus,A. (1948) *J. Immunol.,* **58**, 1.
4. Hammar,J.A. (1938) *Z. Mikrosk. Anat. Forsch.,* **44**, 425.
5. Harris,T.N., Rhoads,J., and Stokes,J. (1948) *J. Immunol.,* **58**, 27.
6. Good,R.A. and Varco,R.L. (1955) *J. Am. Med. Assoc.,* **157**, 713.
7. Miller,J.F.A.P. (1961) *Lancet, ii,* 748.
8. Martinez,C., Kersey,J., Papermaster,B.W., and Good,R.A. (1961) *Proc. Soc. Exp. Biol. Med.,* **109**, 193.
9. Cooper,M.D., Peterson,R.D.A., South,M.A., and Good,R.A. (1966) *J. Exp. Med.,* **123**, 75.
10. Good,R.A. and Gabrielson,E. (1964) *The Thymus in Immunobiology*. Harper Row, New York.
11. Warner,N.L., Szenberg,A., and Burnet,F.M. (1962) *Aust. J. Exp. Biol. Med. Sci.,* **40**, 373.
12. Gowans,J.L. and Knight,E.J. (1964) *Proc. R. Soc., B,* **159**, 257.
13. Auerbach,R. (1961) *Devel. Biol.,* **3**, 336.
14. Hammar,J.A. (1909) *Arch. Mikrosk. Anat. Entw. Mech.,* **73**, 1.

15. Moore,M.A.S. and Owen,J.J.T. (1967) *J. Exp. Med.,* **126**, 715.
16. Owen,J.J.T. and Ritter,M.A. (1969) *J. Exp. Med.,* **129**, 431.
17. Stutman,O., Yunis,E.J., and Good,R.A. (1969) *Transplant. Proc.,* **1**, 614.
18. Claman,H.N., Chaperon,E.A., and Triplett,R.F. (1968) *J. Immunol.,* **97**, 828.
19. Mitchell,G.F. and Miller,J.F.A.P. (1968) *J. Exp. Med.,* **128**, 821.
20. Raff,M.C. (1969) *Nature,* **224**, 378.
21. Knapp,W., Dorken,B., Gilks,W.R., Rieber,E.P., Schmidt,R.E., Stein,H., and von dem Borne,A.E.G.Kr. (ed.) (1990) *Leucocyte Typing IV. White Cell Differentiation Antigens.* Oxford University Press, Oxford.
22. Kisielow,P., Hirst,J.A., Shiku,H., Beverley,P.C.L., Hoffman,M.K., Boyse,E.A., and Oettgen,H.F. (1975) *Nature,* **253**, 219.
23. Rheinherz,E.L., Kung,P.C., Goldstein,G., and Schlossman,S.F. (1979) *Proc. Natl Acad. Sci. USA,* **76**, 4061.
24. Meuer,S.K., Fitzgerald,R., Hussey,J., Schlossman,S., and Reinherz,E.L. (1983) *J. Exp. Med.,* **157**, 705.
25. Hedrick,S.M., Nielsen,E.A., Kavaler,J., Cohen,D.I., and Davis,M.M. (1984) *Nature,* **308**, 153.
26. Yanagi,Y., Yoshikai,Y., Leggett,K., Clark,S.P., Aleksander,I., and Mak,T.W. (1984) *Nature,* **308**, 145,
27. Zinkernagel,R.M. and Doherty,P.C. (1975) *J. Exp. Med.,* **141**, 1427.
28. Burnet,F.M. (1959) *The Clonal Selection Theory of Acquired Immunity.* Cambridge University Press, Cambridge.
29. Kappler,J.W., Roehm,N., and Marrack,P. (1987) *Cell,* **49**, 273.
30. MacDonald,H.R., Schneider,R., Lees,R.K., Howe,R.C., Archa-Orbea,H., Festenstein,H., Zinkernagel,R.M., and Hengartner,H. (1988) *Nature,* **333**, 40.
31. Kieselow,P., Bluthmann,H., Staerz,U.D., Steinmetz,M., and von Boehmer,H. (1988) *Nature,* **333**, 742.
32. Batchelor,J.R., Lombardi,G., and Lechler,R.I. (1989) *Immunol. Today,* **10**, 37.

2

Structure and development

1. Anatomy and basic structure

The adult mammalian thymus is a bilobed structure that lies just above the heart (*Figure 2.1*). Although its position in other vertebrates varies (e.g. in birds the thymus consists of two chains of lobes situated along each side of the neck, while in fish it is within the gill area), the structure of each individual lobe is very comparable in all vertebrate groups. Such phylogenetic differences in position probably represent different degrees of migration of the developing thymic lobes from their initial pharyngeal pouch location (Section 5, this chapter).

Each lobe is surrounded by a connective tissue capsule that at intervals also pushes deep into the tissue, down to the level of the cortico-medullary junction, creating septa that divide the organ into many pseudo-lobules (*Figure 2.2*). These septa carry both the vascular and neuronal supply to and from the thymus, while branches from the septa give rise to the perivascular spaces within the thymus. A basement membrane lies immediately under the capsule and also surrounds the thymic blood vessels. At the histological level, three main thymic areas can be defined: the subcapsular zone that lies, as its name suggests, just under the connective tissue capsule; the cortex, which forms the major outer area; and the centrally located medulla. Each area is characterized by a distinctive lymphoid and stromal cell composition.

The thymus vasculature is supplied by arterioles that enter from the base of the septa in the region of the cortico-medullary junction, and which give rise to three main sets of vessels within the thymus (*Figure 2.3*). Two of these supply the cortex: one forms capillaries that pass out through the cortex to the capsule and then loop around passing back through the cortex, finally leaving via venules at the cortico-medullary junction; the other gives rise to capillaries that pass on out through the cortex and drain into the venous plexus that lies just external to the thymic capsule. The third set of vessels supplies the medulla, forming capillary loops that exit via venules at the cortico-medullary junction. Some of these venules are lined with high endothelium (endothelial cells which are

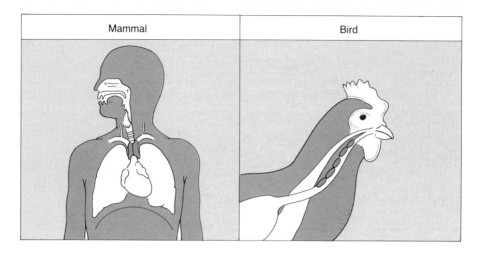

Figure 2.1.   Anatomical location of the thymus in mammals and birds.

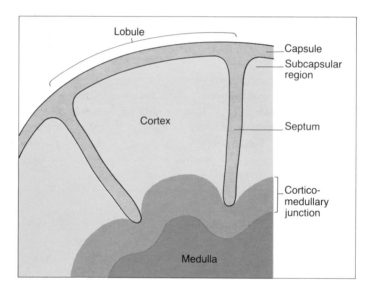

Figure 2.2.   Schematic diagram showing major structural regions of the thymus.

cuboidal rather than flattened—a structure associated with lymphocyte migration). These vessels are termed 'high endothelial venules' and are probably the site of cell traffic into and out of the thymus. A lymphatic system provides drainage for the septa, capsule, and perivascular spaces.

The thymus is innervated from fibres that enter via the septa with the vasculature to form cortico-medullary and subcapsular plexuses (*Figure 2.4*). These are adrenergic and peptidergic. Small cholinergic fibres from the vasculature and cortico-medullary plexus branch throughout the medulla. The

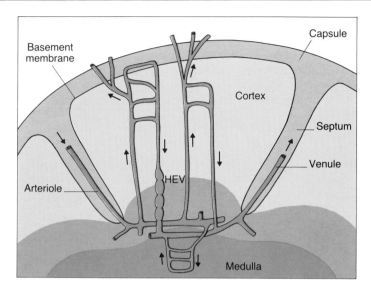

Figure 2.3. Vascular supply to the thymus. Blood vessels enter the thymus via the septa. Capillary loops supply both medulla and cortex; some cortical capillaries pass on through the capsule. High endothelial venules (HEV) are sometimes seen and may be the site of lymphocyte traffic.

cortex is innervated by branching fibres derived from the subcapsular plexus and blood vessels. Recent work is starting to elucidate the peptidergic innervation in the thymus. VIP-like NP-Y and CGRP positive nerve fibres and oxytocin- and vasopressin-containing epithelial cells have been detected by immunostaining techniques.

2. Cellular composition

2.1 Lymphoid cells

The most numerous cells in the thymus are lymphocytes. Approximately 5% of thymic lymphoid cells reside in the subcapsular area; these are mainly large blast cells, reflecting their high mitotic activity. The majority of thymocytes (80–85%) are found in the cortex. Cortical thymocytes are mostly small (smaller than mature T cells), predominantly non-dividing cells that are closely packed together—giving this region of the thymus its characteristic dark histological appearance with nuclear stains (*Figure 2.5*). The remaining 10% of thymocytes are located in the medulla. These lymphocytes are very similar to mature peripheral T lymphocytes in size, phenotype, and functional maturity, and probably represent the end product of intra-thymic development.

2.2 Apoptotic cells

Ninety to 95% of thymocytes are eliminated during T cell receptor repertoire

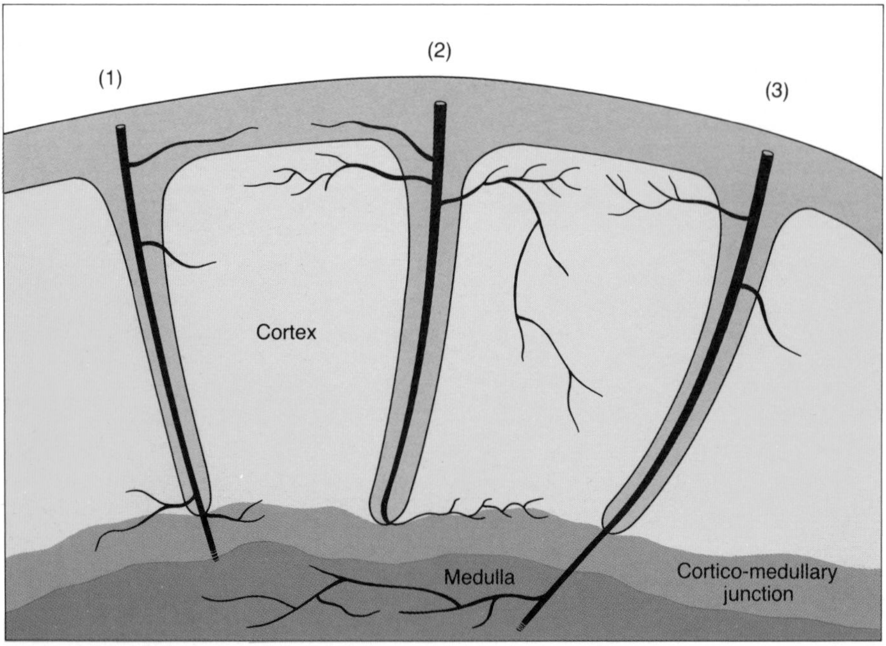

Figure 2.4. Innervation of the thymus. The subcapsular, cortical and cortico‑medullary junction regions of the thymus receive cholinergic, noradrenergic, and peptidergic innervation. Cholinergic innervation is important in the medulla. (1) peptidergic; (2) noradrenergic; (3) cholinergic. Based on ref. 56.

selection (see Chapter 4). Cortical thymocytes are also highly sensitive to external stresses such as acute infection. Death occurs by apoptosis, a programmed suicide mechanism whereby endonucleases cleave nuclear DNA into many smaller pieces (1). Small apoptotic fragments are commonly seen within thymic macrophages (2).

2.3 Stromal cells

Thymic lymphocytes lie within a framework provided by several types of stromal cell (*Figure 2.6*). The major component is the epithelium, which provides the structure within which all the other cell types reside. Epithelial cells in the subcapsular region form a layer one to two cells deep, with the outermost layer lying on the basement membrane; the same cell type lines the septa and perivascular spaces. In rodent thymus the subcapsular layer is less prominent than in man. Epithelial cells in the cortex have very long cytoplasmic processes, forming a network throughout this region of the thymus. In contrast, the epithelium of the medulla consists of oval shaped cells with shorter, spatulate processes that do not form such close connections with each other. Concentric whorls of epithelial cells are also found in the medulla. These Hassall's corpuscles vary considerably in size according to the species, being large in man, rabbit, and guinea pig, but very small in mouse and rat.

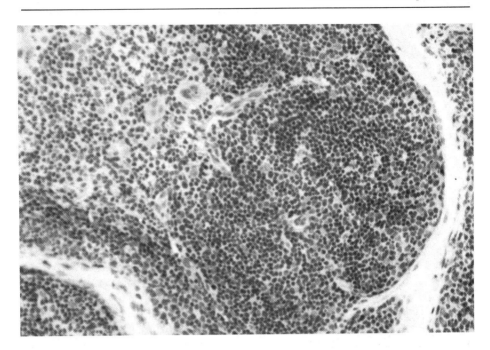

Figure 2.5. Cellular composition of the thymus. Haematoxylin staining of human thymus section to show a lobule with cortical and medullary areas filled with developing thymocytes. Magnification × 431.

Macrophages and dendritic cells (also referred to as interdigitating cells, IDC) form the other major stromal cell populations. Macrophages are scattered throughout both the cortex and medulla; those in the cortex are sometimes surrounded by a rosette of proliferating thymocytes (3). Apoptotic fragments are found predominantly in cortical macrophages, while they are rare in the medulla (2). Dendritic cells are confined to the medulla. Both these cell types are of bone marrow (haemopoietic) origin.

Additional cell types form minority populations within the thymus. Myoid cells have an irregular distribution throughout the thymus; they are most frequently found in clusters in the medulla and are often closely associated with a Hassall's corpuscle. These cells are usually oval/round in shape, although some have tails, and by electron microscopy can be seen to contain skeletal muscle-like striations. They show strong immunostaining with antibodies to actin and striated muscle myosin (4). B lymphocytes, many with an immature phenotype, are found in the medulla, while neutrophils and eosinophils can be seen in many intra-thymic locations and are also prominent in septal areas just outside the thymus. Whether these haemopoietic cells develop within the thymus or whether they represent mature migrants is not known, although the relatively immature phenotype (IgM$^+$) of some of the B cells indicates the former (5,6). Erythropoiesis is sometimes seen, particularly in the avian thymus after periods of stress (as after moulting) (7).

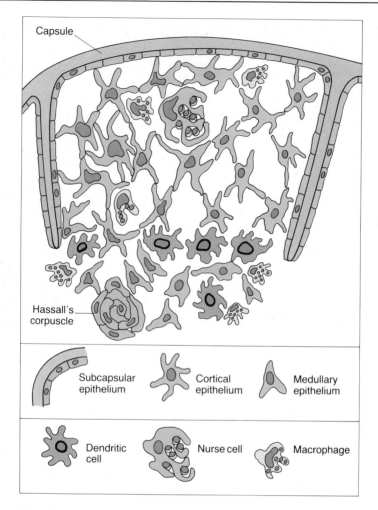

Figure 2.6.   Cellular composition of the thymus. Schematic representation of the stromal cell populations within the thymus. All the intervening spaces are filled with lymphocytes (not shown).

3. Heterogeneity of stromal cells

3.1 Epithelial cells

3.1.1 Keratins

Epithelial cells are characterized by tonofilaments and desmosomes, visualized by electron microscopy, and by the presence of keratin intermediate filaments, which can be detected by immunostaining methods (*Figure 2.7*). Many different keratin sub-types have been identified, and differences in expression of these molecules can be used to distinguish different populations of epithelial cells (8).

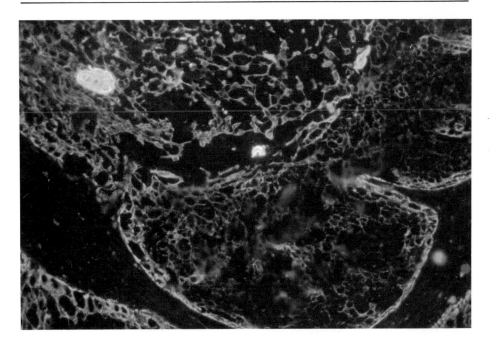

Figure 2.7. Epithelial cells in the thymus. Photomicrograph of a frozen section of human thymus showing immunofluorescence labelling with an anti-keratin antibody, to reveal epithelial cells. Magnification ×431.

Quantitatively, epithelial cells of the subcapsular, perivascular, and medullary regions of the thymus contain considerably more keratin *in toto* than those in the cortex. The most highly keratinized cells are the Hassall's corpuscles. As these concentric whorls develop, the cells in the centre lose their nuclei and closely resemble the keratinocytes in the upper layers of the epidermis (9).

3.1.2 Electron microscopy

Electron microscopy has revealed further heterogeneity within the thymic epithelium (10,11). Six sub-populations (Types 1–6) have been described, according to their EM characteristics. Type 1 cells are the subcapsular epithelial cells, which lie along the surface basement membrane; these cells also line the septa and perivascular spaces. In the cortex it is these Type 1 cells that are thought to contribute towards the 'blood – thymus barrier' (see Section 4, this chapter). Secretory granules are frequently seen within these subcapsular/perivascular cells; at least some of these secretory products are thymic hormones and cytokines (Chapter 5).

Type 2, 3, and 4 cells are found in the cortex and may represent different stages of a single class of epithelial cells, Type 3 being termed 'intermediate', and Type 4 being electron dense distorted dying cells in the deep cortex. Type 2 cells are characterized by their long cytoplasmic processes that extend far from the cell's body, intermesh with those from other Type 2 cells, and surround small islands of cortical thymocytes. This arrangement provides a large surface area for cell–cell interaction between cortical epithelial cells and developing thymocytes. Some of these cortical epithelial cells appear to completely surround and enclose a group of lymphocytes. These structures can be seen, by both EM and light microscopy, in the mid- and outer-cortex and probably represent the *in vivo* equivalent of the thymic nurse cell (TNC) (12,13) (*Figure 2.8*).

TNC were first described after isolation from mouse thymus (14). *In vitro*, these cells completely enclose and contain many small lymphocytes. The number of intra-TNC lymphocytes varies considerably from one or two up to 20–30 in man, while in the mouse TNC are larger and may contain as many as 50. Although *in vitro* these lymphocytes are completely surrounded by TNC membranes (immunostaining studies show that anti-lymphocyte antibodies cannot reach

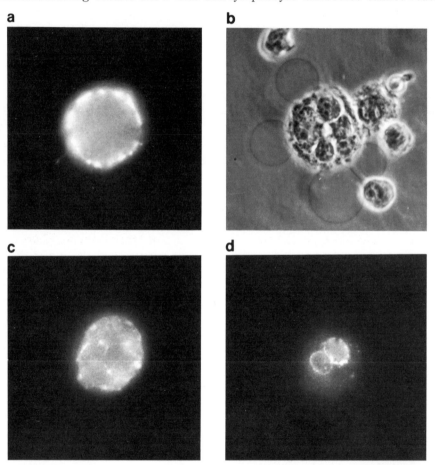

Figure 2.8.   Thymic nurse cells. Immunofluorescence labelling of human TNC in suspension: (a) Thy-1⁺ TNC; (b) the same TNC by phase contrast showing the epithelial cell nucleus and three lymphocytes within the TNC membrane; (c) Thy-1⁺ TNC with permeabilized cell membrane; (d) the same cell after staining with anti-CD1 antibody, showing two cortical thymocytes within the TNC.

them; 12), elegant scanning EM analysis indicates that *in vivo* small gaps exist in the surrounding TNC membranes such that thymocytes can freely enter and exit from these structures (15) (see Chapter 5, *Figure 5.5*). Phenotypic analysis of the intra-TNC lymphocytes shows that they are CD4$^+$,CD8$^+$ 'double positive' cortical thymocytes (Chapter 3).

Two types of medullary epithelial cell can be recognized by electron microscopy. Type 5 cells have relatively short processes, occur in small groups, and do not appear to be active secretory cells. Type 6 cells form a loose network in the medulla, and frequently contain secretory granules. Again, much of this secretory activity is associated with the production of thymic hormones. Type 6 cells form the Hassall's corpuscles. In the fetal human thymus these cells are found scattered singly or in small groups. However, as development proceeds, they form progressively larger concentric whorls of epithelial cells, and those in the centre become highly keratinized, lose their nuclei, and die. In rodent thymus the Hassall's corpuscles remain small throughout life.

3.1.3 Antigenic heterogeneity

A recent approach to the analysis of thymic epithelial cells has been to raise monoclonal antibodies (16,17,18,19), which have revealed considerable molecular heterogeneity within the epithelial component of the thymus. These monoclonal antibodies are currently being characterized and compared in a series of workshops comparable to those used for the human leucocyte reagents (19,20,21,22). Five major 'clusters of thymic epithelial staining' (CTES) have been identified (*Table 2.1*) (*Figures 2.9a, b*). These fit well with the epithelial cell populations defined by light and electron microscopy.

Each epithelial cell subpopulation is distinguished by several different monoclonal antibodies, each recognizing a different target antigen. There are thus many molecular differences between different areas of the thymic epithelium. Such molecular heterogeneity is thought to be associated with functional heterogeneity within the thymic epithelial microenvironment (Chapter 5).

CTES I antibodies are pan-epithelial. They bind to all thymic epithelium and may recognize a shared subtype of keratin. CTES II reagents label subcapsular/perivascular (Type 1) and a major subpopulation of medullary

Table 2.1 Subpopulations of thymic epithelium

Region	EM Type	CTES mab staining	MHC class II expression
Subcapsular/perivascular	1	I, II	−
Cortex	2,3,4,	I, III	+
Nurse cells	2	I, III	+
Medulla	5,6	I, IIp, IV, Vp	+p
Hassall's corpuscles	6	I, II, IV, V	+p

p, subpopulation is positive.

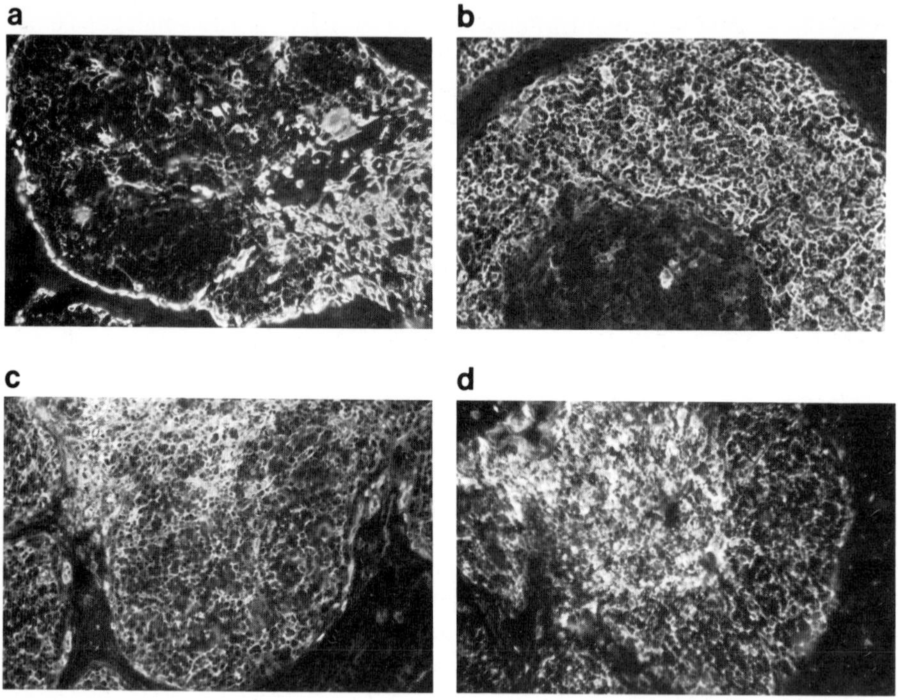

Figure 2.9. Microenvironmental heterogeneity in the thymus. Immunofluorescence labelling of frozen sections of human thymus: (a) CTES II, antibody MR19, 'subcapsule and medulla' epithelial staining pattern; (b) CTES III, antibody MR6, 'cortex' epithelial staining pattern; (c) MHC class I molecules are present on all stromal cells; (d) MHC class II molecules are present on cortical epithelium, medullary dendritic cells, and some medullary epithelium; some thymic macrophages are also MHC class II positive.

(probably Type 6) epithelial cells. Some antibodies in this group recognize thymic hormones, another (A2B5) binds to a ganglioside, while others detect intracellular and cell surface molecules (of unknown structure and function) some of which are also shared by thymic lymphocytes (17,23,24). This epithelium is sometimes referred to as the 'neuroendocrine' epithelium.

CTES III antibodies detect a variety of molecules on cortical epithelial cells, including the TNC (EM Types 2 and 3). The structure and function of these antigens is currently under investigation. Of two characterized so far, one acts as an adhesion molecule while the other appears to be closely associated with the receptor for the lymphokine IL-4 (25).

CTES IV antibodies label Hassall's corpuscles and all medullary epithelial cells (EM Types 5 and 6). CTES V reagents stain Hassall's corpuscles and, sometimes, a surrounding halo of medullary epithelium (EM Type 6).

3.1.4 MHC antigen expression

MHC class I molecules are expressed on the surface of all thymic epithelial cells, at a high and fairly uniform density (*Figure 2.9c*). In contrast, MHC class II molecules are present at high levels on all cortical epithelium, including the TNC, but are completely absent from epithelial cells in the subcapsular region (*Figure 2.9d*). Some medullary cells are also class II positive, but the extent of this medullary labelling varies between different species. In man, relatively few epithelial cells in the medulla express MHC class II molecules; these are mainly found close to the Hassall's corpuscles, although the corpuscles themselves are negative. These class II positive cells are negative with CTES II monoclonal antibodies to subcapsular/medullary epithelium and are probably Type 5 cells. In contrast, in the mouse and rat a larger proportion of medullary epithelial cells bear MHC class II antigens on their surface (26). These molecules are of crucial importance to the intra-thymic selection of the T cell repertoire (see Chapter 4).

3.2 Dendritic cells

A second important stromal cell type in the thymus is the dendritic (or interdigitating; IDC) cell. These cells, which are of bone marrow origin, are located only in the thymic medulla. These are large cells containing a lobulated irregular nucleus with long cytoplasmic projections that extend out amongst the medullary lymphocytes, providing good contacts between the two cell populations. Their cytoplasm contains a 'tubular-vesicular' complex which may be involved in the secretion of soluble factors (3,15).

Limited heterogeneity of dendritic cells has been described. At least some contain Birbeck granules, and hence closely resemble the Langerhans cells of the epidermis. Monoclonal antibody studies have shown that all dendritic cells bear high levels of MHC class II and CD1 molecules; they also weakly expresss the CD4 antigen (27).

3.3 Macrophages

Macrophages are found scattered throughout all thymic areas, but differ in their morphology, and presumably function, in different regions of the thymus. In rodents, three types have been defined: cortical, with enclosed lymphocyte debris, cortico-medullary, with cytoplasmic vacuolar inclusions, and medullary (28). These populations can also be distinguished by differences in their endogenous enzymes, and by their antigenic phenotype defined by monoclonal antibodies (29,30). All macrophages express MHC class I molecules on their surface. In contrast, only 50−75% are MHC class II positive (31,32). Class II expression appears to be a potential property (probably inducible by T cell products) of all subpopulations of thymic macrophages, since class II positive macrophages are found in all areas of the thymus.

3.4 Other stromal components

Other thymic 'stromal' cells such as fibroblasts, vascular endothelium, and myoid cells, together with the extracellular connective tissues, may also be heterogeneous; however, little is known about these thymic components at the present time.

4. Blood – thymus barrier

Type I perivascular epithelial cells surrounding cortical capillaries, together with the tight junctions of the capillary endothelium, are thought to form the structural basis of the 'blood – thymus' barrier thus preventing the entry of macromolecules (such as antigen) into the thymus from the vascular system (33). However, this does not mean that the thymus is sealed from all external molecules. Peripheral antigens have been shown to enter the medullary region (34). Extra-thymic antigens can also gain access to the cortex; serum molecules are exuded from the fenestrated capsular 'arcade' capillaries and subsequently traverse the capsule – cortex boundary together with the flux of interstitial fluid (*Figure 2.10*). These molecules then percolate through the cortex and into the medulla. Thus, the thymocytes can encounter many peripheral self antigens during their intra-thymic development; this may be important in the elimination of auto-reactive cells (negative selection; see Chapter 4) (35,36).

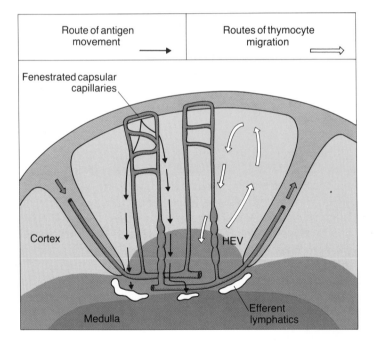

Figure 2.10.   Schematic representation of the flow of antigen through the thymus in relation to the movement of developing T cells. Some antigens may enter the thymus from capsular capillaries and percolate through the cortex to the lymphatics at the cortical medullary junction; developing thymocytes may thus encounter peripheral antigens within the thymus. Based in part on ref. 35.

5. Embryogenesis and the fetal thymus

5.1 Development of the epithelial rudiment

The thymus develops from the third and fourth pharyngeal pouches in mammals, although different pouches are sometimes involved in other animal classes (37). During mammalian development the paired thymic rudiments descend from a cervical location to their final position in the anterior mediastinum, just above the heart. In birds, the thymus remains in the neck, while in fish and amphibia it stays in the pharyngeal/gill region. Despite these differences of location, subsequent development and cellular differentiation are very similar in all classes of animals studied.

Three elements are critical to the normal development of the thymus: ectoderm of the branchial clefts, endoderm of the pharyngeal pouch and mesenchyme from the pharyngeal arch (and ultimately, the neural crest) (38,39,40). If any one of these components is missing the thymus fails to develop. This is well illustrated in the nude mouse, where an absence of the ectodermal component results in failure of the thymus to develop (41). However, there is some controversy as to the exact contribution that each component makes to the final thymic structure. Morphological studies have been interpreted as showing a dual ectodermal – endodermal origin for the thymic epithelium. However, it is difficult to fit this dual origin with the thymic epithelial heterogeneity (Section 3.1). Moreover, recent immunohistological and transplantation studies indicate that there may be a single epithelial stem cell that can give rise to the many different subpopulations of epithelial cells during thymic development. For example: the epithelial cells of the early fetal thymus are phenotypically homogeneous (see later in this section); thymic tumours in both man and mouse are frequently composed of 'double-positive' epithelium, with markers characteristic of both subcapsular and cortical epithelium (CTES II and III); and homogeneous epithelial cells from primary *in vitro* cultures can give rise to all/several normal epithelial subsets when transplanted *in vivo* (42,43,44). The mesenchymal contribution is easier to recognize, since this gives rise to the capsule, septa, and vascular system of the thymus.

The early fetal thymus (chick day 6; mouse day 10; human week 8) is a simple lobular structure surrounded by a connective tissue capsule; there are no septa or major vessels, and it is composed entirely of epithelial cells. It still has its central lumen at this stage. Following this, the lumen becomes compressed and lost due to the expansion of cells within the thymus and the surface becomes indented with mesenchymal septa. These septa ultimately push deep into the thymus (down to the level of the cortico-medullary junction) dividing it into small pseudolobules; they bring with them a vascular and neuronal supply into the thymus.

Initially (e.g. mouse day 10) all epithelial cells belong to an apparently single homogeneous population that share the same morphology and express the same antigenic markers (e.g. all cells are CTES III$^+$). However, soon after this (mouse day 12) differentiation of the epithelium starts with the development of a small island of medullary-type (CTES II$^+$) cells in the centre of the rudiment. By day 14 in the mouse, cortical and medullary epithelium appear to be established with their adult-type phenotype although the elaborate network of the cortical epithelium does not appear until this area is full of lymphocytes. Further developments involve: mitotic division to accommodate the increasing thymic size; the growth and extension of subcapsular epithelium along the septa and around blood vessels, as these elements develop; and the maturation of Hassall's corpuscles, particularly in the human thymus.

5.2 Stem cell entry

Thymocyte precursors are first seen in the thymus soon after development of the epithelial rudiment. At this stage (mouse day 11; chick day 7; man week 9) a small number of large blast cells (the precursors) with dark-staining basophilic cytoplasm can be seen in histological sections. Similar cells are also present in the mesenchymal tissue surrounding the thymus; these can sometimes be seen 'pushing' their way through the thymic capsule and into the thymus (45).

Thymic lymphocyte development is completely dependent upon a supply of these precursor cells (45). Although in the early rudiment they enter through the capsule, as soon as the thymus is fully vascularized they enter via the high endothelial venules in the region of the cortico-medullary junction.

Stem cell entry in mammals (mice) and birds (chicken, quail) occurs in waves, the first coinciding with the initiation of lymphopoiesis in the thymus, with further waves later in embryonic development, each separated by a refractory period

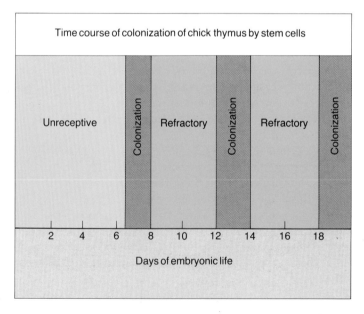

Figure 2.11. Waves of stem cell entry during thymic ontogeny. Periods of colonization, lasting for 1 – 2 days, are separated by intervening refractory periods during embryonic thymus development. Based on refs 46 and 47.

(46,47) (*Figure 2.11*). The inflow of stem cells then continues throughout post-natal and adult life.

Thymic stem cells originate in many different haemopoietic sites, according to the age of the individual. The earliest site to be identified (in birds) is the haemopoietic island surrounding the dorsal aorta (48). At later stages, the yolk sac, fetal liver, and adult bone marrow are the major sources (49).

The potentiality of the stem cells that enter the thymus is not clear. Are these cells pluripotent (i.e. can give rise to all the other haemopoietic lineages)? Are they lymphoid stem cells (committed to either T or B cell lineages)? Or are they already committed to become T lymphocytes? Experimental data suggest that at least some cells that enter the thymus are genuine pluripotent haemopoietic stem cells (50). This is supported by reports of both B lymphocyte and red cell development within the thymus (5,6,7). Other migrants may be progenitor cells that are committed to lymphocyte development (51). Whatever their degree of lineage commitment, a single precursor cell is able to give rise to all the different T cell subpopulations within the developing thymus (52).

5.3 Lymphocyte development

The first thymic lymphocytes appear soon after the initial wave of stem cell entry (mouse day 15; chicken day 9; man week 9). Initially these are scattered throughout the organ, but a typical cortex, tightly packed with small thymocytes, and medulla, more loosely packed with slightly larger thymocytes, are soon clearly seen (mouse day 17, man week 12). The differentiation of these cells in terms of acquisition of typical T cell molecules, rearrangement of T cell receptor genes, and receptor expression is discussed in Chapter 3.

5.4 Bone marrow-derived stromal cells

Macrophages and dendritic cells appear in the developing fetal thymus at about the same time as the first lymphocytes are seen (53). However, although it is known that these cells derive ultimately from a precursor in the bone marrow, it is not clear whether they develop intra-thymically from the pluripotent haemopoietic stem cell population, or whether they enter the thymus as either lineage committed progenitors or mature migrants.

6. Thymic involution and ageing

Two types of thymic involution have been described: acute involution in response to 'stress' and chronic involution that is associated with ageing. The thymus is very sensitive to external 'stress' such as acute infection, malnutrition, surgery, antibiotics and other drugs, and to physiological 'stress' such as pregnancy, lactation, moulting in birds, and metamorphosis in amphibia (54). The effect is mainly mediated via steroid hormones and results in the death of the majority of cortical thymocytes by apoptosis. Thymic regeneration occurs soon after the stress stimulus has been removed.

Age-related involution is characterized by a progressive reduction in thymic size and weight, due to loss of both thymic lymphocytes and stromal cells (epithelium, dendritic cells, and macrophages) (55). The rate of involution is most rapid during the first 10 years of life, after which it continues at a progressively decreasing rate. The total volume of perivascular space and connective tissue increases during the first 20 – 30 years of life, after which the latter is replaced with adipose tissue which comes to form a greater and greater proportion of the organ. However, even in old age (80 – 100 years) small islands of genuine thymic tissue remain. These contain all the individual lymphoid and microenvironmental populations that are seen in the younger thymus. The effect of ageing therefore appears to be predominantly a quantitative one.

The purpose of thymic involution is unknown. However, it is possible that during acute infection, when circulating antigen derived from the invading organism is at a sufficiently high level to reach and filter through the thymus (see Section 4, this chapter), it might provide a protective mechanism against the development of tolerance to the pathogen. In contrast, the chronic involution associated with ageing may partly represent a reduced requirement for T cells in the periphery (once the recirculating pool has become established), and partly a means to reduce the age-associated build-up of mutations that might otherwise increase the risk of autoimmune disease in older age.

7. Further reading

Austyn,J.M. (1989) *Antigen Presenting Cells* (In Focus Series). IRL Press, Oxford.

Haynes,B.F. (1984) The human thymic microenvironment. *Adv. Immunol.*, **36**, 87.

Brekelmans,P. and van Ewijk,W. (1990) Phenotypic characterization of murine microenvironments. *Seminars Immunol.*, **2**, 13.

Kendall,M.D. and Ritter,M.A. (1988) *The Microenvironment of the Human Thymus. Thymus Update* (Annual Review Series), Vol. 1., Harwood Academic, London.

von Gaudecker,B. (1991) Functional histology of the human thymus. *Anat. Embryol.*, **183**, 1.

Muller-Hermelink,H.K. (ed.) The human thymus. *Curr. Topics Pathology*, **75**.

Rouse,R.V. and Weissman,I.L. (1981) Microanatomy of the thymus: its relationship to T cell differentiation. In *Microenvironments in haemopoietic and lymphoid differentiation. Ciba Foundation Symposium*, **84**, 161.

8. References

1. Wyllie,A.H., Morris,R.G., Smith,A.L., and Dunlop,D. (1982) *Am. J. Pathol.*, **109**, 78.

2. Kendall,M.D. (1990) In *The Role of the Thymus in Tolerance Induction,* M.D. Kendall and M.A. Ritter (ed.) *Thymus Update* (Annual Review Series), Vol. 3, p. 47, Harwood Academic, London.

3. von Gaudecker,B. (1991) *Anat. Embryol.*, **183**, 1.

4. Drenckhahn,D., von Gaudecker,B., Muller-Hermelink,H.K., Unsicker,K., and Groschel-Stewart,U. (1979) *Virchows Arch. B.*, **32**, 33.

5. Pabst,R., Binns,R.M., and Westermann,J. (1989) *Thymus*, **13**,149.

6. Isaacson,P.G., Norton,A.J., and Addis,B.J. (1987) *Lancet*, **ii**, 1488.

7. Kendall,M.D. (1984) *Immunol. Today*, **5**, 286.

8. Colic,M., Jovanovic,S., Mitrovic,S.G., and Dujic,A. (1989) *Thymus*, **13**, 175.
9. von Gaudecker,B. and Schmale,E.M. (1974) *Cell Tiss. Res.*, **151**, 347.
10. van de Wijngaert,F.P., Kendall,M.D., Schuurman,H.J., Rademakers,L.H.M.P., and Kater,L. (1985) *Cell Tiss. Res.*, **237**, 227.
11. von Gaudecker,B. (1986) In *The Human Thymus: Histophysiology and Pathology*. Muller-Hermelink,H.K., (ed.), (Current Topics in Pathology, **75**), 1.
12. Ritter,M.A., Sauvage,C.A., and Cotmore,F. (1981) *Immunology*, **44**, 439.
13. van de Wijngaert,F.P., Rademakers,L.H.P.M., Schuurman,H., de Weger,R.A., and Kater,L. (1983) *J. Immunol.*, **130**, 2348.
14. Werkerle,H. and Ketelson,U.P. (1980) *Nature*, **283**, 402.
15. van Ewijk,W. (1988) *Lab. Invest.*, **59**, 579.
16. van Vliet,E., Melis,M., and van Ewijk,W. (1984) *Eur. J. Immunol.*, **14**, 524.
17. Haynes,B.F. (1984) *Adv. Immunol.*, **36**, 87.
18. De Maagd,R.A., Mackenzie,W.A., Schuurman,H.J., Ritter,M.A., Price,K.M., Broekhuizen,R., and Kater,L. (1985) *Immunology*, **54**, 745.
19. Kampinga,J., Berges,S., Boyd,R.L., Brekelmans,P., Colic,M., Van Ewijk,W., Kendall,M.D., Ladyman,H., Nieuwenhuis,P., Ritter,M.A., Schuurman,H., and Tournefier,A. (1989) *Thymus*, **13**, 165.
20. Knapp,W., Dorken,B., Gilks,W.R., Riever,P., Schmidt,R., Stein,H., and van dem Borne,A. (ed.) (1990) *Leucocyte Typing IV. White cell differentiation antigens*. Oxford University Press, Oxford.
21. Ritter,M.A., and Haynes,B.F. (1987) In *Leucocyte Typing III*. McMichael,A. (ed.) Oxford University Press, Oxford, p. 247.
22. Ladyman,H., Boyd,R., Brekelman,P., Colic,M., van Ewijk,W., von Gaudecker,B., Kampinga,J., Kendall,M., Nieuwenhuis,P., Schuurman,H., Tournefier,A., and Ritter,M.A. (1991) In *Lymphatic Tissues and in vivo Immune Responses*. Ezine,S., Berrih-Aknin,S., and Imhof,B. (ed.), Marcel Dekker, New York (in press).
23. Dardenne,M. and Bach,J.F. (1988) In *The Microenvironment of the Human Thymus*. Kendall,M.D. and Harwood,M.A. (ed.), Thymus Update, (Annual Review Series) Vol.1, p.101, Harwood Academic, London.
24. Larche,M., Ladyman,H., and Ritter,M.A. (1987) In *Leucocyte Typing III*. McMichael,A., Beverley,P., Hogg,N., and Horton,M, (ed.), Oxford University Press, Oxford, p. 257.
25. Larche,M., Lamb,J.R., O'Hehir,R.E., Imami-Shita,N., Zanders,E.D., Quint,D.E., Moqbel,R., and Ritter,M.A. (1988) *Immunology*, **65**, 617.
26. Brekelmans,P. and van Ewijk,W. (1990) *Seminars Immunol.*, **2**, 13.
27. Landry,D., Lafontaine,M., Barthelemy,H., Paquette,N., Chartrand,C., Pelletier,M., and Montplaisir,S. (1989) *Eur. J. Immunol.*, **19**, 1855.
28. Milicevic,N.M., Milicevic,Z., Colic,M., and Mujovic,S. (1987) *J. Anat.*, **150**, 89.
29. Dijkstra,C.D., Dopp,E.A., Joling,P., and Kraal,G. (1985) *Immunology*, **54**, 589.
30. Colic,M., Jovanovic,S., Mitrovic,S., and Dujic,A. (1989) *Thymus*, **13**, 175.
31. Nabarra,B. and Papiernik,M. (1988) *Lab. Invest.*, **58**, 524.
32. Beller,D.I. and Unanue,E.R. (1978) *J. Immunol.*, **121**, 1861.
33. Raviola,E. and Karnovsky,M.J. (1972) *J. Exp. Med.*, **136**, 466.
34. Kyewski,B.A., Fathman,C.G., and Rouse,R.V. (1986) *J. Exp. Med.*, **163**, 231.
35. Nieuwenhuis,P. (1990) In *The Role of the Thymus in Tolerance Induction*. Kendall,M.D. and Ritter,M.A. (ed.), *Thymus Update* (Annual Review Series), Vol.3, p.25, Harwood Academic, London.
36. Nieuwenhuis,P., Stet,R.J.M., Wagenaar,J.P.A., Wubbena,A.S., Kampinga,J., and Karrenbeld,A. (1988) *Immunol. Today*, **9**, 372.
37. Lampert,I.A. and Ritter,M.A. (1988) In *The Microenvironment of the Human Thymus*. Kendall,M.D. and Ritter,M.A. (ed.), *Thymus Update* (Annual Review Series), Vol.1, p.5, Harwood Academic, London.
38. Norris,E.H. (1938) *Contribs. Embryol.*, **27**, 191.
39. Cordier,A.C. and Haumont,S.J. (1980) *Am. J. Anat.*, **157**, 227.

40. Salaun,J., Calman,F., Coltrey,M., and Le Douarin,N.M. (1986) *Eur. J. Immunol.,* **16**, 523.
41. Cordier,A.C. and Heremans,J.F. (1975) *Scand. J. Immunol.,* **4**, 193.
42. Willcox,N., Schluep,M., Ritter,M.A., Schuurman,H.J., Newsom-Davis,J. and Christensson,B. (1987) *Am. J. Pathol.,* **127**, 447.
43. Spanopoulou,E., Early,A., Elliott,J., Crispe,N., Ladyman,H., Ritter,M., Watt,S., Grosveld,F., and Kioussis,D. (1989) *Nature,* **342**, 185.
44. Kendall,M.D., Schuurman,H.J., Fenton,J., Broekhuizen,R., and Kampinga,J. (1988) *Cell Tiss. Res.,* **254**, 283.
45. Owen,J.J.T. and Ritter,M.A. (1969) *J. Exp. Med.,* **129**, 431.
46. Jotereau,F.V., Hueze,F., Salomon,V.V., and Gascan,H. (1987) *J. Immunol.,* **138**, 1026.
47. Coltey,M., Jotereau,F.V., and Le Douarin,N.M. (1987) *Cell Differ.,* **22**, 71.
48. Le Douarin,N., Dieterlen-Lievre,F., and Oliver,P.D. (1984) *Am. J. Anat.,* **170**, 261.
49. Moore,M.A.S. and Owen,J.J.T. (1967) *J. Exp. Med.,* **126**, 715.
50. Ezine,S., Papiernik, M., and Lepault,F. (1991) *Int. Immunol.* (in press).
51. Basch,R.S. and Kadish,J.L. (1977) *J. Exp. Med.,* **145**, 405.
52. Williams,G.T., Kingston,R., Owen,M.J., Jenkinson,E.J., and Owen,J.J.T. (1986) *Nature,* **324**, 63.
53. Robinson,J.H. (1984) *Cell. Immunol.,* **84**, 422.
54. Clarke,A.G. and MacLennan,K.A. (1986) *Immunol. Today,* **7**, 204.
55. Steinman,G.G. (1986) In *The Human Thymus,* Muller-Hermelink,H.K. (ed.), (Current Topics in Pathology, **75**), p. 43.
56. Kendall,M.D., Al-Shawaf,A., Aberdeen,J., James,P., and Cowen,T. (1991) Evidence for more than one type of nerve net in the rat thymus gland. In *Lymphatic Tissues and In Vivo Immune Responses.* Imhof,B., Ezine,S., and Berrih-Aknin,S. (ed.), Marcel Dekker, New York (in press).

3

Thymocyte populations and dynamics

This chapter reviews the subpopulations of thymocytes, their origin from bone marrow derived progenitors, their proliferation and differentiation, and their eventual exit to the periphery as functional T cells. During this process, the surface phenotype of these cells evolves through a complex series of stages, which are marked by changes in the expression of a variety of surface antigens. Some of these antigens are molecules of clear functional significance, such as the T cell antigen receptor, CD4, and CD8. Others such as Thy-1, CD5, and CD7 are of unknown function, but may nevertheless be useful as markers which distinguish between subsets (*Table 3.1; Figure 3.1*).

1. Cell sorters

The identification of thymocyte subsets is crucially dependent on multiparameter flow cytometers. These instruments permit large numbers of cells to be individually examined for several physical properties (size, granularity) and for fluorescent signals from ligands bound to the cell surface (antibodies bound to antigens, lectins attached to sugars) or molecules in the cytoplasm (Ca^{2+} dyes). In most flow cytometers, a stream of liquid containing the cells of interest is passed through a laser beam, and scattered laser light and the fluorescent signals emitted are collected by lenses and converted to electrical signals by photodiodes or photomultipliers. Depending on the number of lasers and the sophistication of the signal collection system, current instruments can acquire information on low-angle light scatter, which is related to cell size; right-angle scatter, which reflects cell granularity; and up to four fluorescent markers.

Flow cytometric data for a single parameter (e.g. fluorescein staining) may be displayed as a histogram of the fluorescent signals from each of several thousand cells. For many applications, it is better to display the fluorescent signals on a logarithmic scale, covering three to five decades of signal strength. This has the advantage that the autofluorescence of unstained cells and the signal

Table 3.1. Thymocyte markers

Marker	Synonyms	Mol. Wt (kDa)	Expression pattern in thymus
PNA$^+$	N-Acetyl galactosamine	Various glycoproteins	Cortical thymocytes
Thy-1	Theta	gp17.5	Low on prothymocytes highest on cortical thymocytes intermediate on medullary thymocytes in mouse and rat Prothymocytes and thymic epithelial cells in man
HSA	B2A2,J11d	gp45-55	Precursor and cortical thymocytes Trace level on recent thymus migrants
MHC I	K/D (mouse) HLA-A/B/C	gp44/12	Highest on stem cells and precursors Low on cortical cells Intermediate on medullary cells
TL	Tla	gp44/12	Thymic lymphomas Cortical thymocytes
α/βTCR	Ti, TCR2	gp 38/42	Low on cortical cells High on medullary
γ/δ TCR	TCR1	gp35/44	On fetal and approximately 1% of adult thymocytes
CD1		gp44/12	Cortical thymocytes and medullary dendritic cells
CD2	LFA-2 T11	gp45–50	Thymocytes and T cells, but not the earliest precursors Binds to LFA-3 on thymus epithelial cells
CD3	T3 (human) = p28/gp20/p20/p16 Mouse = gp21/gp28/p25/p16/p21		Associated with α/β and γ/δ TCR Absent from early precursors Low level on cortical cells High level on mature single positive cells
CD4	T4 (human) L3T4 (mouse)	gp55	Expressed on 2/3 of medullary cells Co-expressed with CD8 on cortical cells
CD5	Ly-1 (mouse) Leu-1 (human)	p67	Absent from early precursors Low level on cortical cells High level on mature single-positive cells
CD7		gp40	Human thymocytes and putative bone-marrow prothymocytes
CD8	Ly-2,3 (mouse) T8 (human)	gp38/30 gp34/34	Expressed on 1/3 of medullary cells Co-expressed with CD4 on cortical thymocytes
CD25	IL-2R p55 Tac (human)	p55	Transiently expressed on double-negative cells in mouse
CD44	Pgp-1, Ly-24	gp90	Adhesion molecule on earliest thymocytes
CD45RO	L-CA, T200 Ly-5 (mouse)	gp170–180	Cortical thymocytes

Table 3.1. Continued

Marker	Synonyms	Mol. Wt (kDa)	Expression pattern in thymus
CD45RA	L-CA, T200, Ly-5 (mouse)	gp190, 200 – 205, 220	Medullary thymocytes
MEL-14	Ly-22	gp90	Subsets of CD4⁻,CD8⁻ cells

Summary of the cell surface markers most widely used in the classification of thymocytes. Dimeric or multimeric molecules are indicated with a slash (/) between the weights of the components.
Variable weights due to RNA splicing or glycosylation differences are shown with a dash (–). Established or hypothetical functions of many of these molecules are discussed in this and following chapters.
gp, glycoprotein; p, protein.

from abundant cell surface molecules can be seen on the same scale. Two parameter data are displayed as correlated measurements, most commonly as dot plots or as contour plots. Examining these types of displays, cell subpopulations appear as clusters of high cell density, or as hills standing above a plain. In more fanciful displays, the latter metaphor may be extended by rotating the 'hills' to give an oblique view, with picturesque effect but little gain in clarity.

Flow cytometers also allow the separation of cell populations labelled with different fluorescent markers, or with different levels of a single marker (cell sorting). Cell suspensions are stained, passed through the detection system described above, and a stream of electrically charged droplets is produced with individual cells in some of them. Droplets containing cells of interest may be deflected into a collection tube by a brief pulse of electrostatic charge applied to a deflection plate. Most commercial instruments will reliably sort several thousand cells per second, which places limitations on the number of cells which can be isolated. Nevertheless sorted cells are commonly 99% pure, and have been invaluable in clarifying the roles of different thymic subpopulations.

2. Thymocyte subsets

Most peripheral T cells express either CD4 or CD8, but not both. In contrast, thymocytes express both markers on the majority of cells (1,2) (*Figure 3.1*). Two colour flow cytometric analysis of adult thymocytes for CD4 and CD8 reveals four populations of cells: around 2–4% of the cells are CD4⁻, CD8⁻, approximately 5 and 10% share the CD4⁻,CD8⁺ and CD4⁺,CD8⁻ phenotypes of peripheral T cells, and the vast majority are CD4⁺,CD8⁺. Classification of thymocytes using these markers has been very useful, since the CD4⁻,CD8⁻ subset contains all of the cells with thymus-homing precursor potential (3), while the minority of thymocytes with the functional properties of peripheral T cells are found in the CD4⁻,CD8⁻ and the CD4⁻,CD8⁺ subsets (1,4). The biological role of the major populations of CD4⁺,CD8⁺ cells has been unknown until recently, but it is now clear that a minority of these are available for thymic

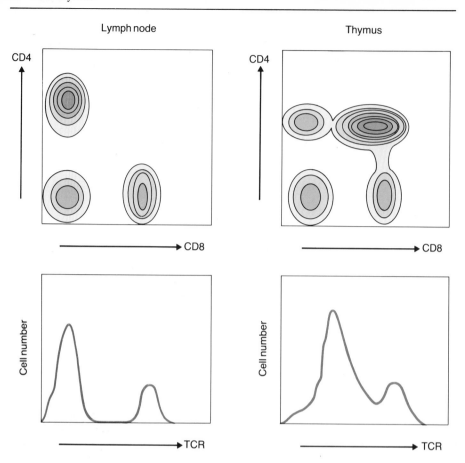

Figure 3.1.   CD4, CD8 and TCR on T-lineage cells. Peripheral lymphoid tissues such as lymph node contain a mixture of cells (e.g. T cells, B cells, macrophages); their T cells express either CD4 or CD8, but not both. These cells express T cell receptors at high density. In contrast, thymocytes are almost all T-lineage cells. Most of them express both CD4 and CD8, while minor subsets are negative for both markers, or are positive for one or the other. Arrows represent increasing intensity of expression.

selection, while the others have already failed selection and are committed to cell death (*Figure 3.2*).

Thymocytes may also be classified using their ability to bind the lectin peanut agglutinin (PNA). In general, PNA-high cells correspond to the non-mature CD4$^-$,CD8$^-$ and CD4$^+$,CD8$^+$ cells which are located in the thymic subcapsular zone and cortex. This fraction of thymocytes also contains the cells which express nuclear terminal deoxynucleotidyl transferase (TdT), an enzyme which has been implicated in the generation of junctional diversity during T cell receptor gene rearrangement (5). In the human, the cell surface marker CD7 is expressed on thymocytes from the very early stages of intra-thymic maturation, and may be useful as an indicator of pre-thymic T-lineage cells (6).

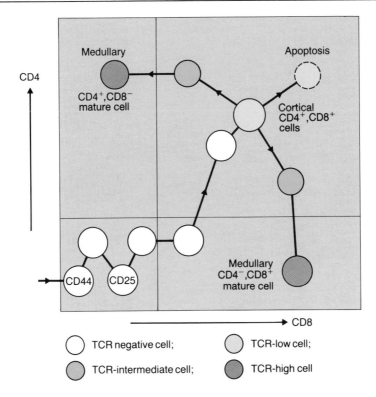

Figure 3.2. Intra-thymic lineage relationships. The earliest thymocytes are CD4⁻,CD8⁻. Within this subset, some very early cells express CD44; a more mature subset express CD25 (the IL-2 receptor p55 chain). Cells pass through a transient CD8⁺ immature stage, then become CD4⁺,CD8⁺, and subsequently express a low level of TCR. These cells are then available for selection. Selected cells are either committed to cell death (apoptosis), or mature through a TCR-intermediate stage and lose the expression of either CD4 or CD8, finally becoming TCR-high medullary cells. Arrows on x and y axes represent increasing intensity of expression.

In the sections which follow, the CD4 and CD8-defined subsets of mouse thymocytes will be described in more detail. The emphasis on the mouse reflects the greater body of experimental work which has been performed with this species.

2.1 Double-negative cells

The term 'double-negative', without further qualification, implies CD4⁻,CD8⁻. Cells of this phenotype, isolated by complement-mediated lysis and/or by cell sorting, will repopulate the irradiated or intact thymus following either intravenous or direct intra-thymic cell transfer (3,7). However, this cell population is heterogeneous, and only some of the constituent subsets have precursor activity (*Figure 3.3*).

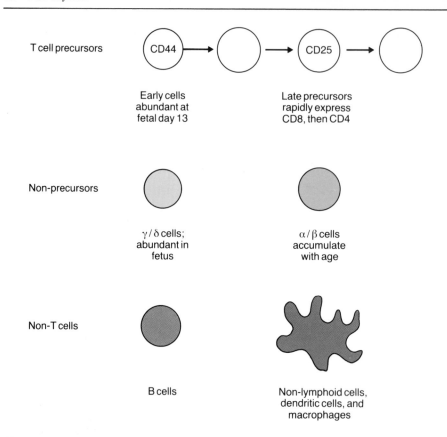

Figure 3.3. Heterogeneity of CD4⁻,CD8⁻ 'thymocytes'. The precursor-enriched CD4⁻,CD8⁻ subset of 'thymocytes' contains at least four early stages of T cell differentiation, and a number of other cell types including γ/δ cells, B cells, and macrophages.

Two major groups of double-negative subsets are defined in the mouse by the antigen HSA (the heat stable antigen, a cell surface glycoprotein widely distributed on haemopoietic cells). The HSA⁺ cells contain essentially all of the precursor activity, most of them express low levels of CD5 (3,8), and a subset of them are the only cells in the thymus to express the p55 (Tac) chain of the receptor for the T cell growth factor interleukin-2 (IL-2R) (7).

The function of IL-2R p55 on these cells is not understood. The earliest HSA⁺,CD4⁻,CD8⁻ cells do not appear to express IL-2R p55, but retain an adhesion molecule CD44 (also known as Pgp-1 or Ly-24) which is common on bone marrow cells (9,10). Later in their differentiation, they express IL-2R p55 transiently before starting to express CD4 or CD8 (7,11). A subset of CD44⁺, HSA-low cells within the HSA⁺,CD4⁻,CD8⁻ cells has been proposed to be the earliest intra-thymic progenitor (11); these cells closely resemble fetal thymocytes at day 13 of gestation (10). In the rat and human, a distinct population of IL-2R p55$^+$ double negative cells has not been described; this contributes to scepticism about the role of the isolated low-affinity IL-2R chain on these cells in the mouse.

The CD4$^-$,CD8$^-$ thymocytes which lack HSA are mostly CD5-high, lack IL-2R, and are devoid of precursor activity. A large part of these cells express antigen receptors of the α/β type, and in some mouse strains their Vβ repertoire is heavily skewed towards the Vβ8 family (12). The function of these cells is not known, but their T cell receptors seem to be functional, and they can be induced to perform some T cell functions *in vitro*, albeit by activation with anti-TCR antibody rather than a physiological ligand (13).

Lymphocytes with CD3-associated γ/δ receptors are also found in the intra-thymic double-negative population (14). Most of these cells are found in the HSA$^+$ subset, and are thus phenotypically similar to the progenitors of α/β cells; however, current evidence suggests that the γ/δ and α/β lineages develop separately within the thymus (15). The use of Vγ gene products by these cells differs dramatically at different ages, which may reflect successive rather than concurrent seeding of various peripheral compartments with γ/δ cells (16).

2.2 Double-positive cells

Thymocytes which express both CD4 and CD8 make up 70–80% of the total. These cells express high levels of HSA and Thy-1, and low levels of MHC class I and α/β antigen receptor (17,18). A minority of these cells are blasts, but most are very small. Double-positives have neither progenitor activity, nor mature T cell function (17), and their place in differentiation has long been problematic. Recently two lines of experimentation have shown that double-positive cells are intermediates between progenitor cells and mature T cells. First, antibodies against CD4 interfere with the intra-thymic shaping of the CD8$^+$ T cell repertoire, which is only explicable if the latter derive from a CD4$^+$ precursor (19,20); second, direct intra-thymic transfer of CD4$^+$,CD8$^+$ blast cells gives rise to minute numbers of more differentiated progeny (21).

2.3 Single-positive CD8$^+$ cells

Cells which express CD8 alone appear very early in ontogeny, at fetal day 15 or 16 (22,23). However these cells are not functionally competent, and correspond to a minor subset of CD8$^+$ cells in the adult thymus. Such cells are HSA$^+$, express little or no T cell receptor, and rapidly become CD4$^+$,CD8$^+$ *in vitro* or *in vivo* (21,24). By all these criteria, the cells seem to be intermediates between double-negative and double-positive cells. This immature population of CD8$^+$ cells have also been described in adult rat thymus, where the cells appear to express a trace of α/β TCR (25,26); and in the developing sheep, where they constitute a major population of thymocytes around day 40 of gestation (27).

The CD8$^+$ single-positive population also contains cells with little or no HSA, but high levels of α/β antigen receptors; these cells account for all of the CD8$^+$ T cell function which can be isolated from the thymus (1).

2.4 Single-positive CD4⁺ cells

2.4 Single-positive CD4⁺ cells

The CD4⁺ single-positive thymocytes are almost all HSA⁻ or HSA-low, and all of them express CD3-associated α/β antigen receptors. These cells are the only population which express the functions of peripheral CD4⁺ T cells (4). Both CD8⁺ and CD4⁺ functionally competent thymocytes are generally held to be at the end of the intra-thymic differentiation process, but it cannot be excluded that some peripheral T cells return to the thymus and appear in this pool (28).

3. Thymocyte population dynamics

Individual bone marrow progenitor cells are able to give rise to very large numbers of thymocytes (29). Around 10–20% of adult thymocytes are blast cells, and a similar proportion are labelled by a single *in vivo* dose of ³H-thymidine, implying that a large fraction of thymocytes are replaced every day. However, measurements of the rate of migration to the periphery suggest only a few per cent are exported (30). The apparent discrepancy is accounted for by intra-thymic cell death. While within the thymus the cells rearrange their T cell receptor genes, express the accessory molecules CD4 and CD8, and are subjected to selective processes based on receptor specificity. These processes may be studied during early ontogeny, by kinetic labelling studies in the steady-state adult, and by direct precursor–product analysis using adoptive transfers.

3.1 Ontogeny

3.1 Ontogeny

Until day 15 of gestation in the mouse, all thymocytes are CD4⁻,CD8⁻. At first these cells are mainly Thy-1⁻, IL-2R⁻ and HSA⁺, CD44⁺, but between days 13 and 15 the expression of CD44 decreases while Thy-1 and IL-2R increase. CD8 is first expressed at day 15 or 16, and CD4 a little later. By day 17–18, the majority of thymocytes are CD4⁺,CD8⁺ as in the adult. Single-positive CD4⁺ or mature CD8⁺ cells do not appear in significant numbers until after birth (10) (*Figure 3.4*). As these phenotypic changes occur, the thymus is increasing rapidly in size, from 1000 cells at day 13 to 10 million at day 18–19. A comparable series of changes occurs in both rat and human thymus (detailed in *Figures 3.4* and *3.5*).

Between days 13 and 15, the mouse thymus is rich in CD3⁺ cells which express γ/δ receptors, the majority of which use the Vγ3 variable region gene (16). A little later in ontogeny, other Vγ gene products predominate, apparently in a time-ordered sequence.

After day 15, the γ/δ cells are overshadowed by a much larger number of cells which express α/β receptors. By fetal day 18, most thymocytes with CD3-associated receptors express α/β receptors at a low surface density, comparable to that on adult double-positive thymocytes.

Some thymocytes at around day 14 express cytoplasmic Vβ chains (and cytoplasmic CD3), but no surface α/β TCR. This is a striking parallel with B

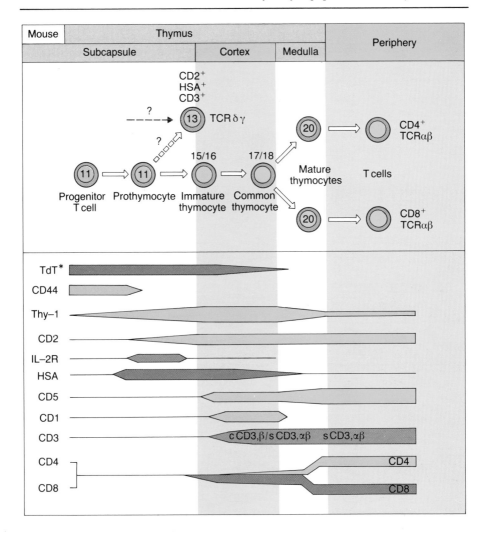

Figure 3.4. A summary of mouse T cell ontogeny, cross-correlating the time of first identification of a T cell subset (in days of gestation, marked inside the cell types) with a panel of cell surface and intracellular markers. The lineage relationship between early γ/δ T cells and α/β cells is unclear. Data for rat thymus are very comparable to those for the mouse, but with the following major differences: early fetal development is delayed by 1–2 days; no immature IL-2R thymocyte population has been observed; although Thy-1 is present on the recent migrants from the thymus it is then lost in the periphery.* TdT+ cells first appear at fetal day 17 in both mouse and rat. This summary is based on data from many sources including refs 8, 10, 16, 25, 44–48. s, surface; c, cytoplasmic.

cell maturation, during which pre-B cells first rearrange their H chain genes and later their L chains; in the interim the cells express cytoplasmic H chains. Another parallel emerges in the ordered rearrangement and expression of $V\gamma$ genes during

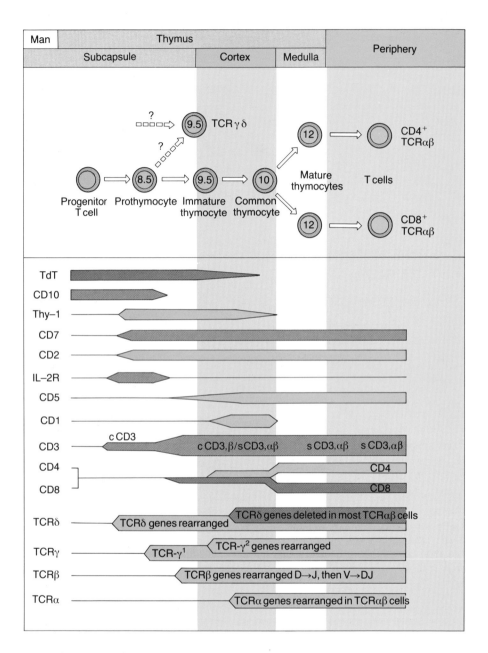

Figure 3.5.   A summary of human T cell ontogeny, cross-correlating the time of first identification of a T cell subset (in weeks of gestation, marked inside the cell types) with a panel of cell surface and intracellular markers and TCR gene rearrangements. The lineage relationship between early α/β T cells and γ/δ cells is unclear. Based on data from many sources including refs 39–41. s, surface; c, cytoplasmic.

ontogeny, which is reminiscent of the sequential rearrangement of progressively more 5' (less C-proximal) VH genes in pre-B cells (31).

3.2 Cell turnover

Progenitor cells competent to home to the thymus can give rise to large numbers of progeny (i.e. they have a large burst size). This suggests a low seeding rate, and experiments with retroviral insertion markers have confirmed that the progeny of a very small number of stem cells are active in the thymus at any one time (29,32). This implies a high rate of intra-thymic cell division, and abundant dividing cells are found in the thymus. The cycling cells are found mainly in the HSA$^+$,CD4$^-$,CD8$^-$ cells, the immature HSA$^+$,CD4$^-$,CD8$^+$ cells, and the large CD4$^+$,CD8$^+$ cells. *In vivo* pulse labelling with ^{3}H-thymidine or with bromodeoxyuridine suggests that the cycling populations give rise to the non-cycling small CD4$^+$,CD8$^+$ cells and the mature CD4$^+$ or CD8$^+$ single-positive cells within a few days (17,33).

3.3 Precursor – product relationships

The differentiation pathway from the earliest precursor cells to mature primary T cells has been clarified in rodents by reconstitution experiments, in which highly purified subpopulations of thymocytes have been injected directly into the thymus, and their differentiation products have been identified and characterized by multiparameter flow cytometry (*Figure 3.4*). The earliest cells appear to be CD4$^-$,CD8$^-$ cells which express HSA and CD44 (11). These cells are engaged in rearranging the T cell receptor β locus. After several days, the cells transiently express the p55 chain of the IL-2R (7,11), and subsequently begin to express CD8. In the rat, α/β T cell receptors are expressed at this stage (26); in the mouse they are probably not expressed until later (34).

Immature CD8$^+$ cells very rapidly express CD4 and become double-positive blast cells. At this stage, they have already completed rearrangement of the β locus, and may express β chains in the cytoplasm. After successful rearrangement of the α locus, α/β T cell receptors are expressed. At around this time, the cells stop dividing, and become available for selective processes (35). The cells then have three fates. The majority become small double-positive cells, which die probably without leaving the thymus [although an alternative model suggests their final resting place is elsewhere (36)]. A minority become CD4$^+$ or CD8$^+$ single-positive cells, which exit to the periphery (21). Identification of recent thymus migrants is possible following injection of FITC (fluorescein isothiocyanate) into the thymus. Fluoresceinated cells identified in the periphery shortly after this procedure are phenotypically and functionally very similar to most other T cells, except that they express a trace of the HSA marker (30).

A similar pathway of T lymphocyte differentiation has been inferred for man, using phenotypic analysis of normal thymocytes, and leukaemic cells where normally rare cell subpopulations have become expanded and are available for study (*Figure 3.5*).

4. Changes in cell location

In the early fetus, histological studies suggest that progenitor cells migrate from the surrounding mesenchyme and enter the thymus rudiment through the developing capsule. However in more mature animals, these cells arrive via the blood and are most likely extracted from the circulation by high endothelial venules (HEV), which are anatomically similar to those in lymph nodes and are found at the cortico-medullary junction. However, the greatest density of dividing cells in the thymus is found in the subcapsular zone of the outer cortex, and it is reasonable to assume that many of the early developmental steps occur there (37). Dipping experiments, in which intact thymi are immersed briefly in FITC to surface-fluoresceinate outer cortical cells, suggest that most $CD4^-,CD8^-$ cells and the immature subset of $CD4^-,CD8^+$ cells are subcapsular in location.

Later $CD4^+,CD8^+$ cells are found throughout the cortex, and selection of self-MHC restriction probably occurs mainly in this phase of differentiation. While cortical thymocytes are mainly double-negative or double-positive, $CD4^+$ and mature $CD8^+$ single positive cells are located mainly in the medulla. It is therefore likely that maturing cells pass back into the medulla before they are permitted to leave the thymus, and the location of a dense network of dendritic cells at the cortico-medullary junction can be interpreted as a filter in which cells are screened for self-reactivity before the acquisition of functional competence (38).

Finally, a small number of peripheral T, and B, lymphocytes do recirculate back into the thymic medulla where they constitute a small but persistent population (49). Such cells might influence thymocyte maturation and selection processes.

5. Further reading

Fowlkes,B.J. and Pardoll,D.M. (1989) Molecular and cellular events of T cell development. *Adv. Immunol.,* **44**, 207.

Haynes,B.F., Denning,S.M., Singer,K.H., and Kurtzberg,J. (1989) Ontogeny of T-cell precursors: a model for the initial stages of human T-cell development. *Immunol. Today,* **10**, 87.

Janossy,G., Campana,D., and Akbar,A. (1989) Kinetics of T lymphocyte development. In *Cell Kinetics of the Inflammatory Response.* Iverson,O.H. (ed.), *Curr. Topics Pathol.,* **79**, 59.

Kendall,M.D. and Ritter,M.A. (eds) (1989) *T Lymphocyte Differentiation in the Human Thymus.* Thymus Update (Annual Review Series), Vol. 2, Harwood Academic, London.

Rothenberg,E. (1990) Death and transfiguration of cortical thymocytes: a reconsideration. *Immunol. Today,* **11**, 116.

Jenkinson,E.J. and Owen,J.J.T. (1990) *Seminars Immunol.,* **2**, 51.

6. References

1. Ceredig,R., Glasebrook,A.L., and MacDonald,H.R. (1982) *J. Exp. Med.,* **155**, 358.
2. Dialynas,D.P., Quan,Z.S., Wall,K.A., Pierres,A., Quintans,J., Loken,M.R., *et al.* (1983) *J. Immunol.,* **131**, 2445.

3. Fowlkes,B.J., Edison,L., Mathieson,B.J., and Chused,T.M. (1985) *J. Exp. Med.*, **162**, 802.
4. Ceredig,R., Dialynas,D.P., Fitch,F.W., and MacDonald,H.R. (1983) *J. Exp. Med.*, **158**, 1654.
5. Rothenberg,E. (1980) *Cell*, **20**, 1.
6. Haynes,B.F., Martin,M.E., Kay,H.H., and Kurtzberg,J. (1988) *J. Exp. Med.*, **168**, 1061.
7. Shimonkevitz,R.P., Husmann,L.A., Bevan,M.J., and Crispe,I.N. (1987) *Nature*, **329**, 157.
8. Crispe,I.N., Moore,M.W., Husmann,L.A., Smith,L., Bevan,M.J., and Shimonkevitz,R.P. (1987) *Nature*, **329**, 336.
9. Hyman,R., Lesley,J., Schulte,R., and Trotter,J. (1986) *Cell. Immunol.*, **101**, 320.
10. Husmann,L.A., Shimonkevitz,R.P., Crispe,I.N., and Bevan,M.J. (1988) *J. Immunol.*, **141**, 736.
11. Pearse,M., Li,W., Egerton,M., Wilson,A., Shortman,K., and Scollay,R. (1989) *Proc. Natl Acad. Sci. USA*, **86**, 1614.
12. Budd,R.C., Miescher,G.C., Howe,R.C., Lees,R.K., Bron,C., and MacDonald,H.R. (1987) *J. Exp. Med.*, **166**, 577.
13. Fowlkes,B.J., Kruisbeek,A.M., Ton-That,H., Weston,M.A., Coligan,J.E., Schwartz,R.H., and Pardoll,D.M. (1987) *Nature*, **329**, 251.
14. Lew,A.M., Pardoll,D.M., Maloy,W.L., Fowlkes,B.J., Kruisbeek,A., Cheng,S.-F., *et al.* (1986) *Science*, **234**, 1401.
15. Winoto,A. and Baltimore,D. (1989) *Nature*, **338**, 430.
16. Havran,W.L. and Allison,J.P. (1988) *Nature*, **335**, 443.
17. Scollay,R., Bartlett,P., and Shortman,K. (1984) *Immunol. Rev.*, **82**, 79.
18. Roehm,N., Herron,L., Cambier,J., DiGuisto,D., Kappler,J., and Marrack,P. (1984) *Cell*, **38**, 577.
19. Fowlkes,B.J., Schwartz,R.H., and Pardoll,D.M. (1988) *Nature*, **334**, 620.
20. Macdonald,H.R., Hengartner,H., and Pedrazzini,T. (1988) *Nature*, **335**, 174.
21. Guidos,C.J., Weissman,I.L., and Adkins,B. (1989) *Proc. Natl Acad. Sci. USA*, **86**, 7542.
22. Ceredig,R., MacDonald,H.R., and Jenkinson,E.J. (1983) *Eur. J. Immunol.*, **13**, 185.
23. Kisielow,P., Lieserson,W., and von Boehmer,H. (1984) *J. Immunol.*, **133**, 1117.
24. MacDonald,H.R., Budd,R.C., and Howe,R.C. (1988) *Eur. J. Immunol.*, **18**, 519.
25. Paterson,D.J. and Williams,A.F. (1987) *J. Exp. Med.*, **166**, 1603.
26. Hunig,T., Wallny,H.-J., Hartley,J.K., Lawetzky,A., and Tiefenthaler,G. (1989) *J. Exp. Med.*, **169**, 73.
27. Mackay,C.R., Maddox,J.F., and Brandon,M.R. (1986) *J. Immunol.*, **136**, 1592.
28. Fink,P.J., Bevan,M.J., and Weissman,I.L. (1984) *J. Exp. Med.*, **159**, 436.
29. Ezine,S., Weissman,I.L., and Rouse,R.V. (1984) *Nature*, **309**, 629.
30. Scollay,R., Chen,W.-F., and Shortman,K. (1984) *J. Immunol.*, **132**, 25.
31. Alt,F., Blackwell,K., and Yancopoulos,G.D. (1987) *Science*, **238**, 1511.
32. Snodgrass,R. and Keller,G. (1987) *EMBO J.*, **6**, 3955.
33. Penit,C. (1986) *J. Immunol.*, **137**, 2155.
34. Bluestone,J.A., Pardoll,D., Sharrow,S.O., and Fowlkes,B.J. (1987) *Nature*, **326**, 82.
35. Richie,E.R., McEntire,B., Crispe,N., Kimura,J., Lanier,L.L., and Allison,J.P. (1988) *Proc. Natl Acad. Sci. USA*, **85**, 1174.
36. Rothenberg,E. (1990) *Immunol. Today*, **11**, 116.
37. Hirokawa,K., Sado,T., Kubo,S., Kamisaku,H., Kitomi,K., and Utsuyama,M. (1985) *J. Immunol.*, **134**, 3615.
38. Farr,A.G., Anderson,S.K., Marrack,P., and Kappler,J. (1985) *Cell*, **43**, 543.
39. Ritter,M.A., Sauvage,C.A., and Delia,D. (1983) *Immunology*, **49**, 555.
40. Greaves,M.F., Hariri,G., Newman,R.A., Sutherland,D.R., Ritter,M.A., and Ritz,J. (1983) *Blood*, **61**, 628.
41. Lobach,D.F., Hensley,L.L., Ho,W., and Haynes,B.F. (1985) *J. Immunol.*, **135**, 1752.

42. Campana,D., Janossy,G., Coustan-Smith,E., Amlot,P., Tian,W.T., Ip,S., and Wong,L. (1989) *J. Immunol.,* **142**, 57.
43. van Dongen,J.J.M., Comans-Bitter,W.M., Wolvers-Tettero,I.L.M., and Borst,J. (1990) *Thymus,* **16**, 207.
44. Kampinga,J. and Aspinall,R. (1990) In Kendall,M.D. and Ritter,M.A. *The Role of the Thymus in Tolerance Induction.* (Annual Review Series), Vol. 3, p. 149, Harwood Academic, London.
45. Fowlkes,B.J. and Pardoll,D.M. (1989) *Adv. Immunol.,* **44**, 207.
46. Owen,J.J.T., Kingston,R., and Jenkinson,E.J. (1986) *Immunol.,* **59**, 23.
47. Ritter,M.A., Gordon,L.K., and Goldschneider,I. (1978) *J. Immunol.,* **121**, 2463.
48. Whittum,J., Goldschneider,I., Greiner,D., and Zurier,R. (1985) *J. Immunol.,* **135**, 272.
49. Michie,S.A. and Rouse,R.V. (1989) *Thymus,* **13**, 141.

Note added in proof: It has recently been shown that a very early population of intra-thymic progenitor cells expresses low amounts of CD4 (Wu *et al.* (1991) *Nature,* **349**, 71–74). This population of cells has the TCR β and γ genes in germline configuration, and gives rise to both $\alpha\beta$ and $\gamma\delta$ cells in adoptive transfer. Thus these cells are likely to be earlier than CD4$^-$,CD8$^-$ progenitor cells, and may include cells not yet committed to the $\alpha\beta$ or the $\gamma\delta$ lineage. The differences between these CD4-dull progenitor cells and bone marrow multipotential haemopoietic stem cells (Spangrude *et al.* (1989) *Science,* **241**, 58–62) have not yet been clarified.

4

The T cell repertoire

Progenitor cells enter the thymus without a functional antigen receptor, and exit either with α/β or with γ/δ receptors. While the functions and specificity of γ/δ T cells are still under investigation, α/β T cells emerge from the thymus with well-defined biological properties. They recognize exogenous antigens associated with class I and class II molecules of the major histocompatibility complex, and they are normally functionally tolerant of self-components. This chapter outlines current views on the generation of the α/β and γ/δ T cell specificity repertoires, and on the modification of these repertoires which occurs during differentiation in the thymus.

1. Generation of the repertoire

T cell receptors are encoded in complex genetic loci which undergo rearrangement to produce functional genes (1). These genes were originally identified on the basis of their rearrangement (2), and they share, with B cell immunoglobulin genes, not only the use of rearrangement to generate diversity, but also some components of the same DNA recombinase enzyme system (3).

During B and T lymphocyte receptor gene recombination, for one chain (Ig, light chain; TCR, α or γ) one of many domain-sized variable (V) regions is juxtaposed to one of several short joining (J) gene segments; while for the other chain (Ig, heavy chain; TCR, β or δ) an additional short diversity (D) segment is interposed between the V and J gene segments. This generates combinatorial diversity. The potential variability of Ig and TCR proteins is enhanced by the imperfect joining of recombining gene segments, with the loss or gain (under the action of the nuclear enzyme TdT) of nucleotides (1). This superimposes junctional diversity on the combinatorial diversity. Detailed accounts of the structural organization and rearrangement mechanisms of these loci may be found in many reviews (e.g. *Immune Recognition* by J.R.Lamb and M.J.Owen, IRL Press, In Focus Series) and only the main points relevant to T cell ontogeny are outlined here (*Figures 4.1* and *4.2*).

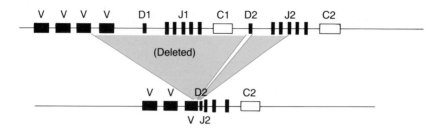

Murine TCR β locus in germline configuration

(Deleted)

Rearranged β locus in T cells

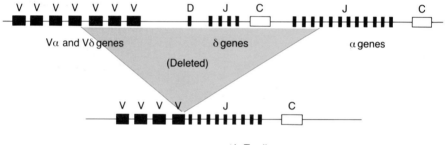

α / δ locus in germline configuration

Vα and Vδ genes   δ genes   α genes

(Deleted)

α rearrangement in T cells

Figure 4.1.   Rearrangements of the β and α/δ loci which occur in α/β T cells. The β locus forms a VDJ join with two DNA segments excised, while α rearrangements of the α/δ locus form a VJ join and excise one very large segment of DNA containing all of the delta sequences.

The β and γ genes are encoded on chromosomes 6 and 13 in the mouse. These loci contain small numbers of variable (V) genes, perhaps 20–25 for β and less than 10 for γ (4,5), separated from several constant region genes by joining (J) and, for β, diversity (D) segments. Gene rearrangment follows the patterns established for B cell Ig genes, with D–J rearrangement preceding V–D joining. Early models of the lineage relationships between γ/δ and α/β cells suggested that γ, and its partner δ, might be the first to rearrange, and that the fate of the cell might depend on the success or otherwise of this rearrangement (6). However it now seems that γ, β and δ all rearrange very early in T cell ontogeny (7), although it is not clear whether this process starts intra-thymically, or can also occur before precursor cells reach the thymus.

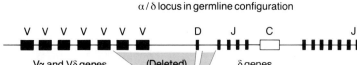

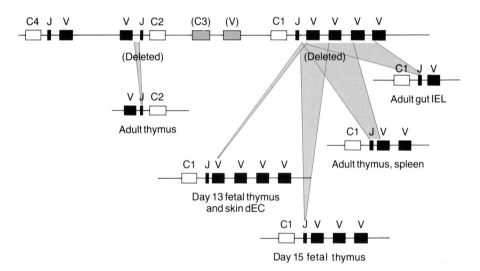

Figure 4.2. Rearrangements of the α/δ and γ loci which occur in γ/δ cells. In this case the V regions of the α/δ locus undergo a VDJ rearrangement, deleting two gene segments. The various Vγ genes undergo VJ rearrangements which are transcribed with one of several C genes. The pattern of expression of the rearranged genes is tightly restricted in time and tissue distribution.

The α and δ T cell receptor genes are encoded in a complex locus, on chromosome 14 in the mouse. The D, J, and single C regions of δ are nested between the Vα/Vδ gene cluster and an extensive array of Jα segments. There is some overlap between the V regions which rearrange to form δ and α. Alpha (α), like δ, has a single C region. The structure of the locus requires that the δ segment be looped out and deleted before rearrangement of α is possible, and productive rearrangements of α are not found at the earliest times in ontogeny

(8); α rearrangement seems to be limiting for α/β T cell receptor expression, and receptors of this type are not seen until day 16 of ontogeny.

The clonally invariant CD3 components associated with both types of T cell receptor do not depend on gene rearrangement, and are probably the first parts of the T cell receptor to be expressed. In man, cytoplasmic CD3 staining (using antibody to the CD3 ϵ chain) can be found in the fetal liver as early as 10 weeks of gestation, around the time that colonization of the thymus begins (9). This suggests that when precursor cells reach the thymus they may already express cytoplasmic CD3, and that T cell receptor assembly and cell surface expression is limited by the rearranging genes.

2. Recognition of antigen by T cells

The α/β T cell receptor interacts with antigen presented as a complex with class I or class II molecules encoded by the major histocompatibility complex (MHC). MHC molecules are highly polymorphic, and the allelic variation in these molecules as well as the sequence of the antigenic epitope determines T cell recognition specificity. Thus T cells are specific for a particular antigen fragment associated with a particular MHC molecule. As a consequence, antigen-specific T cells from one individual generally will not recognize their appropriate antigen presented by cells from a genetically dissimilar member of the same species. This property of T cell recognition is termed MHC restriction (10 – 13).

2.1 T cell receptor interactions with MHC

Conventional α/β cells are functionally tolerant of self-molecules, and recognize exogenous antigens associated with self-alleles of the MHC. Both CD4$^+$, MHC class II restricted T cells and CD8$^+$, MHC class I restricted T cells appear to recognize antigens in extended conformation, in most cases as peptide fragments of the antigenic protein (14). The recently determined X-ray crystallographic structure of HLA-A2 is compatible with the idea that antigenic peptides are presented by a specialized region of the MHC class I molecule, flanked by two helical strands of the MHC chain, one from the $\alpha-1$ and one from the α-2 homology unit (15,16). Self-tolerance therefore implies unresponsiveness to self-peptides presented by self-MHC molecules, while MHC restriction implies that T cells are able to recognize exogenous antigens presented by self-MHC molecules, rather than by non-MHC self-molecules. In both of these situations, the thymus is centrally concerned in the definition of self.

It has been proposed, on the basis of sequence similarity, that T cell receptor α and β chains fold and combine like the H and L chains of immunoglobulin (1). If this predicted structure of the T cell receptor is 'docked' with the known crystallographic structure of the class I MHC, the V-gene encoded parts of the receptor may be juxtaposed to the flanking α-helical parts of the MHC molecule, while the central VD and DJ junctional regions of the T cell receptor make contact with the antigenic peptide (*Figure 4.3*). From this model, it has been inferred

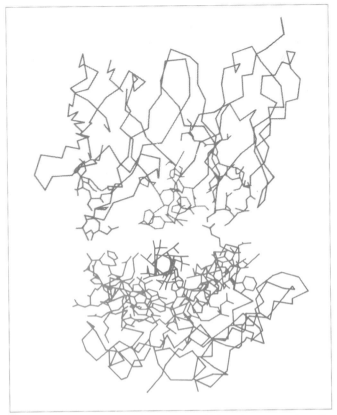

Figure 4.3.   Interaction between TCR and MHC-antigen complex. Representation of the structure of a hypothetical TCR (brown) interacting with an antigenic peptide (dark orange) held in α-helical configuration in the cleft of an HLA-A2 molecule (orange) as seen from the side. Redrawn from ref. 1.

that V gene diversity parallels the polymorphism of MHC genes, while junctional diversity creates ligands for the world of antigenic peptides.

2.2 Self-tolerance

T cells are functionally tolerant to self-MHC molecules, to ubiquitous non-MHC self-molecules, and to self-molecules expressed at anatomical sites remote from the thymus. Three general classes of mechanisms have been proposed to account for these properties, namely clonal deletion (17), clonal anergy without deletion (18), and active suppression (19). Before the mid-1980s, it was difficult to distinguish experimentally between these possibilities, because potentially self-reactive T cells could not be identified except by tests of T cell function. More recently, a set of monoclonal antibodies against specific T cell receptor Vβ regions have been used to detect clonal deletion directly. Where an identifiable T cell receptor Vβ has significant reactivity to a self-antigen in association with a self-MHC molecule, the following general rules apply. Thymocytes bearing the particular $V\beta$ region are present at almost normal levels in the CD4$^+$,CD8$^+$, CD3-low population, but are deleted (or greatly reduced) from mature thymocytes and from peripheral T cell populations (20) (*Figure 4.4*). In most of the cases studied, deletion affects both the CD4$^+$ and the CD8$^+$ T cells, probably because deletion occurs at the CD4$^+$,CD8$^+$ stage of differentiation (20,21). Histologically, the V region is present at normal levels in the thymic cortex, but reduced or absent from the medulla.

In T cell receptor transgenic mice created by the injection of fertilized mouse ova with rearranged α and β chain genes, deletion appears to occur at an earlier stage of maturation, as soon as low levels of CD8 are expressed (22) (*Figure 4.5*). In these animals, the timing of T cell receptor gene expression is disturbed because there is no delay in α chain synthesis. This argues that the mechanism of clonal deletion is not subject to precise developmental timing; instead, it can start as soon as α/βTCR and CD8 are co-expressed.

Not all cells of the thymus are equally capable of imposing tolerance on developing T cells. In radiation chimeras, bone marrow-derived cells are effective in inducing tolerance (23), while isolated thymus epithelium is not (24). Developing T cells seem able to tolerize one another, but only as far as class I MHC responses are concerned (25). In transgenic mice which express the $V\beta17a$ gene, and H-2E molecules only on the thymus cortex (epithelium), T cell

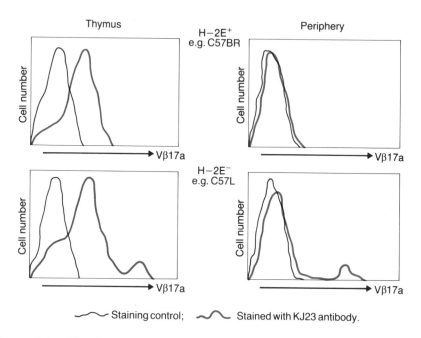

Figure 4.4. Clonal deletion of $V\beta17a$. A number of T cell receptor $V\beta$ genes have been discovered which confer reactivity for H-2E molecules. Such T cells are deleted in H-2E-expressing mouse strains. In the prototype system, T cells with $V\beta17a$ are expressed in the periphery and as TCR-high thymocytes only in H-2E-negative strains. Arrow represents increasing intensity of $V\beta17a$ expression.

receptors using this H-2E-reactive V region are not deleted. However, when the same H-2E gene is expressed selectively in the medulla (mainly bone marrow-derived cells), deletion occurs (26). These observations support a widely-held view that the majority of clonal deletion occurs at the cortico-medullary junction or in the medulla.

The direct demonstration of clonal deletion of a growing list of $V\beta$ regions by different self-molecules does not imply that this phenomenon accounts for all of self-tolerance. On the contrary, in several experimental systems, self-reactive T cell receptors evidently co-exist in apparently stable equilibrium with their antigens. Such anergy is evident in Mls-1[b] mice injected with foreign cells expressing the Mls-1[a] gene; here $V\beta6$ T cells are not deleted, but

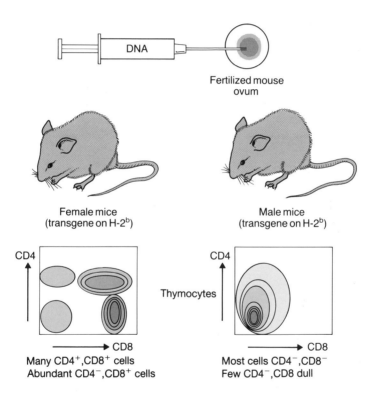

Figure 4.5. Clonal deletion of T cells which recognize a non-MHC self-antigen. Rearranged TCR α and β genes were isolated from a T cell clone which recognizes the male-specific transplantation antigen H-Y in the context of H-2D[b]. DNA constructs were injected into mouse ova, to create transgenic mice. In H-2[b] transgenic females, the thymus contained all of the normal CD4 and CD8 defined cell populations, although CD8[+] single positive cells outnumbered CD4[+] single positives. In transgenic males, the thymus was small and most of the thymocytes were CD4[-],CD8[-]. This was interpreted as clonal deletion, occurring early in the CD4[+],CD8[+] phase of maturation. Arrows represent increasing intensity of expression.

responsiveness to Mls-1[a] is abolished (27). Several *in vitro* systems display similar behaviour (18). It may be that this mechanism complements intra-thymic clonal deletion, allowing self-tolerance to self-molecules which are unable to gain access to the thymus. At present it is not clear whether clonal deletion and clonal anergy are two quite distinct processes, or whether they are different outcomes of the same transmembrane signalling events, differing only in the maturity of the responding cell.

Thus within the thymus it is likely that most cells will be tolerized by deletional events, while in the periphery clonal anergy is thought to play a major role in maintaining self-tolerance. For a long time it was thought that the blood – thymus barrier prevented access of peripheral antigens to the developing cortical lymphocytes, and that tolerance to all peripheral antigens must occur in the periphery. However, recent experimental data suggest that peripheral antigens can reach both the thymic medulla (28) and the cortex (29) (see also Chapter 1), although in the latter case there is a size limit (macromolecules of 150 kDa can gain access while those of 950 kDa cannot).

A further mechanism, 'T suppression', has also been invoked to explain peripheral tolerance. However, despite 15 years of active study, the role of active suppression in self-tolerance remains highly controversial. Some research is based on the concept that there exists a distinct subset of T cells with suppressive activity (19), but there are alternative explanations for apparent suppressive effects, such as cytotoxicity directed against syngeneic effector cells (30), and inappropriate antigen presentation, for instance by other T cells (31) (*Figure 4.6*). It seems likely that these suppressor effects will ultimately be explicable in terms of the cytotoxic and lymphokine secreting activities of CD4[+] and CD8[+] T cells, rather than by the existence of a distinct subset of suppressor T cells.

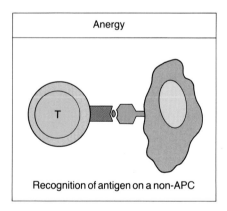
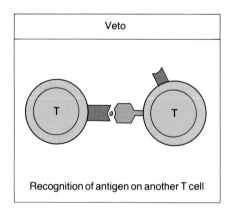

Figure 4.6.   T cells may be inactivated in the periphery by recognition of antigen on cell types which are inefficient at antigen presentation, possibly because these cells lack the capacity to deliver an unidentified 'second signal' to the T cell. Anergy and veto may be two names for the same phenomenon.

2.3 T cell restriction specificity

When bone marrow precursor cells heterozygous for the MHC are allowed to differentiate in an irradiated thymus of one parental haplotype, the restriction specificity of the T cells which develop is strongly skewed towards the MHC of the thymus donor (32,33). Grafting experiments using fetal thymus lobes which have been depleted of haemopoietic cells (lymphocytes, macrophages, dendritic cells) by culture in 2-deoxyguanosine (34) have shown that the tissue responsible for control of T cell restriction specificity is the thymus epithelium (35) (*Figure 4.7*). To explain this phenomenon, it was proposed that differentiating thymocytes are tested for their restriction specificity, and those potentially restricted to self-MHC molecules are selectively permitted to mature. This is termed positive selection.

For many years the positive selection model was disputed, partly because of

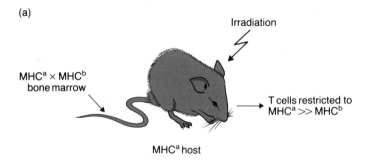

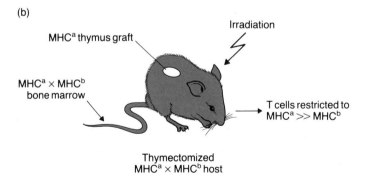

Figure 4.7. The control of T cell restriction specificity in chimeric animals. Irradiated mice reconstituted with semi-allogenic F1 bone marrow developed T cells restricted to the MHC of the radioresistant host (a). In irradiated, thymectomized, bone-marrow reconstituted F1 animals which had been grafted with either an intact or an irradiated thymus, T cell restriction was controlled by the thymus genotype (b). In refinements of the second type of experiment, the grafts were deoxyguanosine-treated (haemopoietic cell depleted) fetal thymus lobes; here T cell restriction was controlled by the thymus lobe genotype.

anomalous results obtained in different types of chimera experiments (36), and partly because of the theoretical difficulty of understanding how the same T cell receptor could be selected both for and against self-reactivity. As in the case of clonal deletion, alternative explanations could be devised for experiments based on the analysis of expressed T cell specificity. For instance, in an irradiated thymectomized (P1 × P2)F1 mouse grafted with a parental P1 thymus and F1 bone marrow, the lack of P2-restricted reactivity of F1 T cells which had matured in the P1 parental thymus might be due to P2-specific suppressor cells (36). A clear picture of positive selection has emerged from studies of mice transgenic for T cell receptor genes of defined specificity. In the most extensively

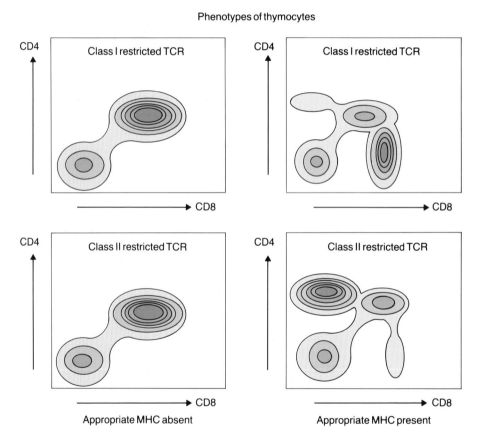

Figure 4.8. Positive selection in mice which express transgenic class I or class II restricted antigen receptors. In TCR-transgenic mice which lack the appropriate MHC molecules, T cell maturation halts at the CD4+,CD8+ stage. If an MHC class I restricted transgenic TCR encounters its restriction element, the T cells rapidly exit from the CD4+,CD8+ subset and become CD8+. Similarly, transgenic MHC class II restricted receptors cause diversion of T cells into the CD4+ subset in the presence of their restriction element. Arrows represent increasing intensity of expression.

characterized set of such mice, T cell receptors specific for the male transplantation antigen H-Y in the context of H-2D^b are preferentially expressed in CD8$^+$ T cells (22) (*Figure 4.8*). In radiation chimeras, the genotype of the host rather than that of the transgenic bone marrow determines the expression of the transgene (37). When the transgenes were back-crossed onto SCID mice, essentially all of the T cells followed the rules of positive selection (in this case, T cells with an MHC class I restricted TCR develop towards the CD8$^+$ single positive subset), and in the absence of the H-2D^b restriction element T cell differentiation was arrested at the CD4$^+$,CD8$^+$ stage (38). Similar effects have been observed in other transgenic mice expressing MHC class I restricted receptors (39). Conversely, where mice are transgenic for an MHC class II restricted TCR, T cell differentiation is directed towards the CD4$^+$ mature single positive subpopulation (40) (*Figure 4.8*).

In normal mice, several Vβ regions identified by monoclonal antibodies appear to be controlled by positive selection processes. In these cases, high expression of the particular Vβ in CD4$^+$ T cells was dependent on the MHC haplotype of the mouse, and in F1 mice high expression was dominant (41,42). Thus positive selection functions in normal, polyclonal mice as well as in transgenic animals of limited clonal diversity and there seems no reason to entertain other explanations of the control of T cell restriction specificity by the thymus.

Although positive selection has been convincingly demonstrated, the cellular and molecular mechanisms responsible are still obscure. To explain how a particular MHC molecule can positively select T cells so as to bias their restriction specificity, while another can cause their deletion, several models have been proposed. One possibility is that the two processes act at different stages of differentiation. Another, non-exclusive idea is that positive selection requires only a very low receptor affinity, while clonal deletion acts at higher affinities. The net result would be the survival of cells with a low but non-zero affinity for self-MHC, some of which would display higher cross-reactive affinity for exogenous antigens presented by self-MHC. An idea gaining in stature is the proposal that thymic epithelial MHC is different from all other MHC, probably because it presents a largely non-overlapping spectrum of peptides which mimic the world of exogenous antigens (43). In this concept, positive selection occurs on an internal image of possible antigens, and distinctive affinity rules are not required. This risky theory implies that thymus epithelium is permanently antigenic, but survives because this tissue is very inefficient at inducing immune responses in unprimed T cells (24).

2.4 Ontogeny and specificity of γ/δ cells

The fetal mouse thymus between days 13 and 15 of gestation contains CD3$^+$ cells which express on the cell surface γ/δ receptors, most or all of which use a distinctive Vγ gene (*Figure 4.2*). In the adult, intra-thymic γ/δ cells employ different Vγ genes, and the early fetal predominant gene is found only in dendritic epidermal (dEC) cells of the skin (44). This has provoked the suggestion that a first wave of γ/δ cells is committed to home to the epidermis, while subsequent waves are destined for the gut, other epithelia, and the major lymphoid organs. Strikingly, γ/δ cells from several sources use a different predominant Vγ gene at each anatomical site. In the skin, this uniformity extends to Vδ use and to junctional diversity (44), while in the gut intra-epithelial lymphocytes have somewhat more diversity (45). Overall, the available data reveal much less receptor variability in γ/δ than in α/β T cells.

The limited diversity of γ/δ receptors, coupled with their competence to perform several T cell functions (46), supports the idea that they are defensive cells which recognize not antigens expressed by pathogens, but a limited range of host-encoded proteins induced by various stresses (45). One category of such host-encoded proteins are heat-shock proteins, which are synthesized in response to a raised temperature but also to virus infection, chemical toxicity, or mechanical stimuli such as the contact of macrophages with plastic. Some γ/δ cells respond to mycobacterial antigens (47), and a large component of the response is directed to heat-shock proteins, some of which are cross-reactive with host proteins expressed on infected or otherwise damaged host cells.

In terms of the Davis–Bjorkman model (1) which assigns MHC specificity to V regions and antigen specificity to junctional segments of the TCR molecule, the limited V region diversity of γ/δ receptors might suggest that they respond to antigens in the context of MHC-like molecules that are less polymorphic than conventional class I or class II. In line with this idea, several human and murine γ/δ T cell clones have been described which respond to non-classical MHC-like molecules such as Qa-1, TL and CD1 (48–50). This area of research is very active and other γ/δ ligands are likely to be proposed.

2.5 Selection of γ/δ cells

The growing perception that γ/δ cells function similarly to α/β T cells in their recognition of appropriately presented antigens has prompted consideration of whether they are subject to repertoire selection, either for self-tolerance or for self-restriction. Since antigen-specific immune responses with *ex vivo* γ/δ cells have been difficult to demonstrate, and a systematic survey of Vγ and Vδ expression in human families or in different mouse strains has not been reported, most of the evidence on this point comes from transgenic mice.

Two sets of γ/δ TCR transgenic mice provide evidence for tolerance mechanisms which resemble those of α/β T cells. In one case, γ and δ transgenes from a class I (TL^a)-reactive clone were apparently deleted in mice which expressed the antigen; the deletion was dominant in F1 mice, supporting clonal deletion as the mechanism (51). In another case, transgenic TL^b-reactive γ/δ receptors were expressed on peripheral cells, but the cells were functionally inactivated in TL^b mice; as with many instances of α/β T cell anergy, the cells regained the response to antigen in the presence of exogenous IL-2 (52). From the limited data available, it seems that γ/δ cells are subject to both deletion and anergy mechanisms of self-tolerance, in a very similar way to α/β T cells.

Evidence for positive selection of γ/δ cells is more sketchy, but the presence of γ/δ cells in mice with defective MHC class I expression due to inactivation of the β_2-microglobulin genes (53) suggests that a normal level of conventional class I expression is not essential for γ/δ cell maturation. However, it is not certain that all Qa/Tla class I molecules absolutely require β_2-microglobulin for their expression, and the question of positive selection of γ/δ cells remains unresolved.

3. Further reading

Austyn,J.M. (1989) *Antigen-presenting Cells* (In Focus series). IRL Press, Oxford.

Blackman,M., Kappler,J., and Marrack,P. (1990) The role of the T cell receptor in positive and negative selection of developing T cells. *Science*, **248**, 1335.

Burkly,L.C., Lo,D., and Flavell,R.A. (1990) Tolerance in transgenic mice expressing major histocompatibility molecules extrathymically on pancreatic cells. *Science*, **248**, 1364.

Jenkinson,E.J. and Owen,J.J.T. (1990) *Seminars Immunol.*, **2**, 51.

Kendall,M.D. and Ritter,M.A. (ed.) (1990) *The Role of the Thymus in Tolerance Induction. Thymus Update* (Annual Review Series), Vol. 3, Harwood Academic, London.

Owen,M.J. and Lamb,J.R. (1988) *Immune Recognition* (In Focus series). IRL Press, Oxford.

Ramsdell,F. and Fowlkes,B.J. (1990) The role of the thymus in inducing self tolerance. *Science*, **248**, 1342.

Rothenberg,E.V. (1990) Death and transfiguration of cortical thymocytes: a reconsideration. *Immunol. Today*, **11**, 116.

Sprent,J., Lo,D., Gao,E.K., and Ron,Y. (1988) T cell selection in the thymus. *Immunol. Rev.*, **101**, 173.

Sprent,J., Gao,E.K., and Webb,S.T. (1990) T cell reactivity to MHC molecules: immunity versus tolerance. *Science*, **248**, 1357.

von Boehmer, H., Teh,H.S., and Kisielow,P. (1989) The thymus selects the useful, neglects the useless and destroys the harmful. *Immunol. Today*, **10**, 57.

von Boehmer, H. and Kisielow, P. (1990) Self – nonself discrimination by T cells. *Science*, **248**, 1369.

4. References

1. Davis,M.M. and Bjorkman,P.J. (1988) *Nature*, **334**, 402.
2. Chien,Y.-H., Gascoigne,N.J.R., Kavaler,J., Lee,N.E., and Davis,M.M. (1984) *Nature*, **309**, 322.
3. Schuler,W., Weiler,I.J., Schuler,A., Phillips,R.A., Rosenberg,N., Mak,T.W., *et al.* (1986) *Cell*, **46**, 963.
4. Behlke,M.A., Spinella,D.G., Chou,H.S., Sha,W., Hartl,D.L., and Loh,D.Y. (1985) *Science*, **229**, 566.
5. Strominger,J.L. (1989) *Cell*, **57**, 895.
6. Pardoll,D.M., Fowlkes,B.J., Bluestone,J.A., Kruisbeek,A., Maloy,W.L., Coligan,J.E., and Schwartz,R.H. (1987) *Nature*, **326**, 79.
7. Winoto,A. and Baltimore,D. (1989) *Nature*, **338**, 430.
8. Haars,R., Kronenberg,M., Gallatin,M.W., Weissman,I.L., Owen,F.L., and Hood,L. (1986) *J. Exp. Med.*, **164**, 1.
9. Campana,D., Janossy,G., Coustan-Smith,E., Amlot,P.L., Tian,W.-T., Ip,S., and Wong,L. (1989) *J. Immunol.*, **142**, 57.
10. Rosenthal,A. and Shevach,E. (1973) *J. Exp. Med.*, **138**, 1194.
11. Zinkernagel,R.M. and Doherty,P.H. (1974) *Nature*, **248**, 701.
12. Bevan,M.J. (1975) *J. Exp. Med.*, **142**, 1349.

13. Sprent,J., Korngold,R., and Molnar-Kimber,K. (1980) *Springer Sem. Immunopathol.*, **3**, 213.
14. Townsend,A. and Bodmer,H. (1989) *Ann. Rev. Immunol.*, **7**, 601.
15. Bjorkman,P.J., Saper,M.A., Samraoui,B., Bennett,W.S., Strominger,J.L., and Wiley,D.C. (1987) *Nature*, **329**, 506.
16. Bjorkman,P.J., Saper,M.A., Samraoui,B., Bennett,W.S., Strominger,J.L., and Wiley,D.C. (1987) *Nature*, **329**, 512.
17. Good,M.F., Pyke,K.W., and Nossal,G.J.V. (1983) *Proc. Natl Acad. Sci. USA*, **80**, 3045.
18. Mueller,D.L., Jenkins,M.K., and Schwartz,R.H. (1989) *Ann. Rev. Immunol.*, **7**, 445.
19. Cunningham,A.J. (1976) *Transplantation*, **31**, 23.
20. Kappler,J., Roehm,N., and Marrack,P. (1987) *Cell*, **49**, 273.
21. MacDonald,H.R., Schneider,R., Lees,R.K., Howe,R.C., Acha-Orbea,H., Festenstein,H., *et al.* (1988) *Nature*, **332**, 40.
22. Kisielow,P., Bluthmann,H., Staerz,U.D., Steinmetz,M., and von Boehmer,H. (1988) *Nature*, **333**, 742.
23. Sprent,J., von Boehmer,H., and Nabholz,M. (1975) *J. Exp. Med.*, **142**, 321.
24. von Boehmer,H. and Schubiger,R. (1984) *Eur. J. Immunol.*, **14**, 1084.
25. Shimonkevitz,R.P. and Bevan,M.J. (1988) *J. Exp. Med.*, **168**, 143.
26. Marrack,P., Lo,D., Brinster,R., Palmiter,R., Burkly,L., Flavell,R.H., and Kappler,J. (1988) *Cell*, **53**, 627.
27. Rammensee,H.-G., Kroschewski,R., and Frangoulis,B. (1989) *Nature*, **339**, 541.
28. Kyewski,B.A., Fathman,C.G., and Rouse,R.V. (1986) *J. Exp. Med.*, **163**, 231.
29. Nieuwenhuis,P., Stet,R.J.M., Wagenaar,J.P.A., Wubbena,A.S., Kampinga,J., and Karrenbeld,A. (1988) *Immunol. Today*, **9**, 372.
30. Shinohara,N., Watanabe,M., Sachs,D.H., and Hozumi,N. (1988) *Nature*, **336**, 481.
31. Fink,P.J., Shimonkevitz,R.P., and Bevan,M.J. (1988) *Ann. Rev. Immunol.*, **6**, 115.
32. Bevan,M. and Fink,P. (1978) *Immunol. Rev.*, **42**, 3.
33. Zinkernagel,R.M., Callahan,G.N., Althage,A., Cooper,S., Klein,P.A., and Klein,J. (1976) *J. Exp. Med.*, **147**, 892.
34. Jenkinson,E.J., Franchi,L.L., Kingston,R., and Owen,J.J.T. (1982) *Eur. J. Immunol.*, **12**, 583.
35. Lo,D. and Sprent,J. (1986) *Nature*, **319**, 672.
36. Matzinger,P. and Mirkwood,G. (1978) *J. Exp. Med.*, **148**, 84.
37. Kisielow,P., Teh,H.-S., Bluthmann,H., and von Boehmer,H. (1988) *Nature*, **335**, 730.
38. Scott,B., Bluthmann,H., Teh,H.S., and von Boehmer,H. (1989) *Nature*, **338**, 591.
39. Sha,W.C., Nelson,C.A., Newberry,R.D., Kranz,D.M., Russell,J.H., and Loh,D.Y. (1988) *Nature*, **336**, 73.
40. Berg,L., Pullen,A.M., Fazekas de St. Groth,B., Mathis,D., Benoist,C., and Davis,M. (1989) *Cell*, **58**, 1035.
41. MacDonald,H.R., Lees,R., Schneider,R., Zinkernagel,R.M., and Hengartner,H. (1988) *Nature*, **336**, 471.
42. Blackman,M.A., Marrack,P., and Kappler,J. (1989) *Science*, **244**, 214.
43. Marrack,P., McCormack,J., and Kappler,J. (1989) *Nature*, **338**, 503.
44. Asarnow,D.M., Kuziel,W.A., Bonyhadi,M., Tigelaar,R.E., Tucker,P.W., and Allison,J.P. (1988) *Cell*, **55**, 837.
45. Takagaki,Y., DeCloux,A., Bonneville,M., and Tonegawa,S. (1989) *Nature*, **339**, 712.
46. Cron,R.Q., Gajewski,T.F., Sharrow,S.O., Fitch,F.W., Martin,L.A., and Bluestone,J.A. (1989) *J. Immunol.*, **142**, 3754.
47. Modlin,R.L., Pirmez,C., Hofman,F., Torigian,V., Uyemura,K., Rea,T.H., *et al.* (1989) *Nature*, **339**, 544.
48. Vidovic,D., Roglic,M., McKune,K., Guerder,S., MacKay,C., and Dembic,Z. (1989) *Nature*, **340**, 646.
49. Porcelli,S., Brenner,M.B., Greenstein,J.L., Balk,S.P., Terhorst,C., and Belcher,P.A.

(1989) *Nature*, **341**, 447.

50. Ito,K., Van Kaer,L., Bonneville,M., Hsu,S., Murphy,D.P., and Tonegawa,S. (1990) *Cell*, **62**, 549.

51. Dent,A.L., Matis,L.A., Hooshmand,F., Widacki,S.M., Bluestone,J.A., and Hedrick,S.M. (1990) *Nature*, **545**, 714.

52. Bonneville,M., Ishida,L., Itohara,S., Verbeek,S., Berns,A., Kanagawa,O., Haas,W., and Tonegawa,S. (1990) *Nature*, **344**, 163.

53. Zijlstra,M., Bix,M., Simister,N.E., Loring,J.M., Raulet,D.H., and Jaenisch,R. (1990) *Nature*, **344**, 742.

5

The thymic microenvironment

1. Introduction

Normal T lymphocyte development is dependent upon the presence of the thymus, and involves both direct contact with thymic stroma as well as interaction with secreted soluble molecules (1,2). Collectively, the cells, extracellular connective tissue, and soluble molecules that provide this unique inductive role of the thymus are termed the 'thymic microenvironment'. The full extent of this microenvironment is not known; however, it is likely that in addition to the recognized contribution of epithelial cells, macrophages and dendritic cells (and their soluble products), other components such as neurones, extracellular connective tissue—and even the lymphocytes themselves—may also play an important part. Developing T cells require signals for: entry, positioning and migration within the thymus, proliferation, maturation, T cell receptor gene rearrangement and expression, positive and negative selection, and emigration to the periphery. In this chapter, the cell – cell, cell – soluble molecule interactions involved in these processes will be described as far as they are known. This is, however, an area of very active research and there are many gaps in the story. The current picture should therefore be regarded as the 'tip of the iceberg'.

2. Cell – cell interactions

2.1 The in vivo situation

Two main sources of material can be used to analyse cell – cell interactions in the thymus. Data gained from light and electron microscope studies of thymus tissue sections can provide information on the relative positioning of different cell types *in situ,* and hence which cell populations have the potential to interact with each other (*Figures 5.1* and *5.2*). Immunoelectron microscopy studies, showing the precise cellular location of antigens (cell surface, intracellular compartment), can extend this to the molecular level (*Figures 5.3* and *5.4*).

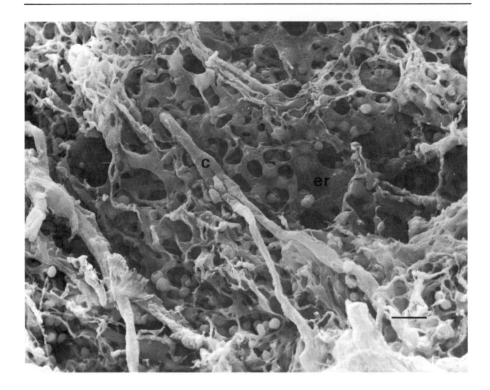

Figure 5.1.   Thymic cortex revealing the general structure of the epithelial network (er). 'Arcade' capillaries (c) run through this network up to the capsule, where they loop back towards the medulla (not shown). Thymocytes are situated in between the spider shaped epithelial cells. Scanning electron microscopy; mouse thymus, bar indicates 20 μm. Photograph courtesy of W.van Ewijk, from ref. 6 with permission.

2.1.1 Subcapsular zone

After entry into the thymus, the thymocyte precursors (CD4$^-$,CD8$^-$ double-negative prothymocytes) localize to the subcapsular region. At this stage of their development the stromal cells with which they can potentially interact are the subcapsular epithelial cells and scattered macrophages. CD8 and CD4 immature single-positive thymocytes also occupy this area of the thymus.

2.1.2 Cortical zone

Double-positive, CD4$^+$,CD8$^+$, thymocytes are found in the cortex where they can interact with the cortical epithelial cells (including the thymic nurse cells), perivascular epithelium, and again, scattered macrophages (*Figures 5.1, 5.3, 5.4 and 5.5*).

The presence of macrophage – lymphocyte rosettes in this area, with one macrophage surrounded be several thymocytes in mitosis (3), indicates an important role for macrophages in the stimulation of early thymocyte

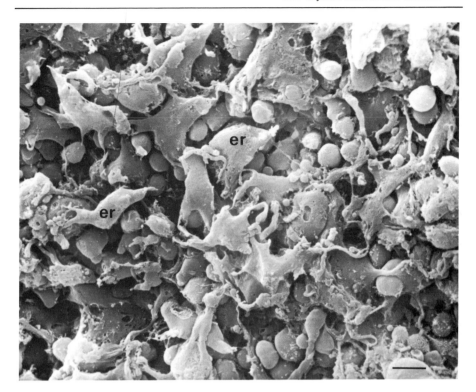

Figure 5.2.   Thymic medulla. Epithelial cells (er) are densely packed, leaving little extracellular space. They have only short cytoplasmic extensions. Scanning electron microscopy, mouse thymus; bar indicates 10 μm. Photograph courtesy of W.van Ewijk, from ref. 6 with permission.

proliferation. Other macrophages in this area are filled with apoptotic debris (4), presumably representing the deletion of auto-reactive thymocytes (see Chapter 4). Whether this negative selection has been induced by direct interaction between developing thymocytes and the macrophages, or whether these macrophages are simply acting as bystander phagocytic cells is not known.

The importance of thymocyte – epithelial interactions in the cortex is highlighted by the almost complete engulfment of cortical thymocytes by the TNC (5,6), and by the observation that the T cell receptor molecules on some of the thymocytes are polarized to their points of contact with the cortical thymic epithelial cells (7) (*Figures 5.4, 5.5,* and *5.6*). Transplantation experiments using thymus grafts depleted of all lymphocytes, macrophages, and dendritic cells have shown that MHC restriction of developing lymphocytes is induced by a thymic epithelial cell (8).

Further studies using transgenic mice, where expression of H-2E MHC class II molecules was directed to different stromal cell types, have shown that it is the cortical epithelium that is critical in the positive selection of MHC restricted T cells (9) (see Chapter 4).

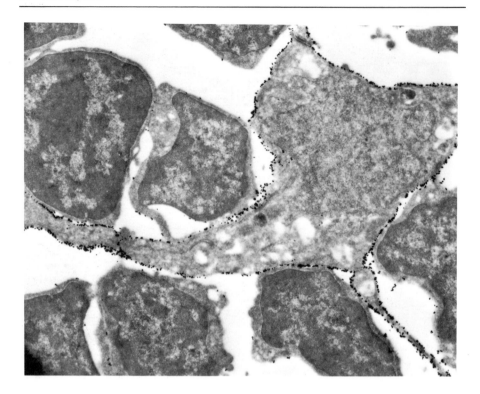

Figure 5.3.   A cortical thymic epithelial cell labelled with CTES III antibody MR6. Human thymus, immunogold electron microscopy, magnification ×8100. Photograph courtesy of B.von Gaudecker, from ref. 47 with permission.

2.1.3 Medullary zone

Mature, single positive, CD4⁺ or CD8⁺ thymocytes are located in the medullary region of the thymus where they can potentially interact with macrophages, dendritic cells, and two different populations of epithelial cell (Types 5 and 6) (including the Hassall's corpuscle) (*Figure 5.2*). Electron microscopy shows close interactions between lymphocytes and the stromal cell populations, particularly with the surface projections of the dendritic cells. Thymus grafting and transgenic mouse experiments demonstrate that it is a bone marrow derived cell, most probably the medullary dendritic cell, that is of major importance in negative selection especially of class II reactive cells (10,11) (see Chapter 4). However, it is possible that class II positive macrophages, scattered throughout the thymus, and the small populations of medullary B cells may also be involved in this process (12).

2.2 In vitro *studies*

The sequential association of developing thymocytes with different populations of stromal cells has been studied experimentally using gently dissociated thymic

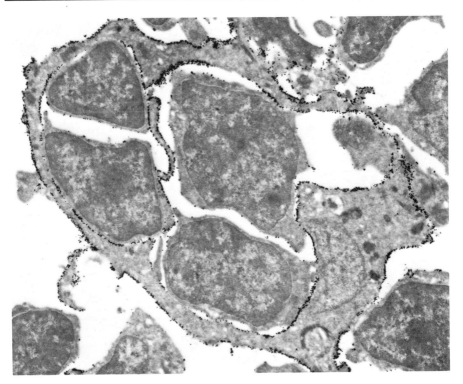

Figure 5.4.   A thymic nurse cell (TNC) labelled with CTES III antibody MR6. Four lymphocytes can be seen within the TNC; human thymus, immunogold electron microscopy, magnification ×8100. Photograph courtesy of B.von Gaudecker, from ref. 47 with permission.

tissue, removed at a series of time points after entry of antigenically marked lymphocyte precursors (13,14). Double-positive (CD4$^+$,CD8$^+$) thymocytes were first found associated with macrophages, then within TNC, and finally clustered around dendritic cells. This suggests that once they have reached the double-positive stage the developing T lymphocytes first receive a proliferation stimulus (macrophage) followed by positive (TNC) and negative (dendritic cell) selection. However, the cellular associations of the earlier double-negative and immature single positive CD4$^+$ and CD8$^+$ thymocytes are unknown. An additional, antigen independent, proliferative stimulus for developing T cells may be provided by epithelial cells (*Figure 5.7*). Primary cultures of thymic epithelium (subset unknown) induce thymocyte proliferation via an interaction between CD2 molecules on the surface of the thymocytes and LFA3 on the epithelium (15). A further role of the epithelium may be to induce lymphocyte differentiation (*Figure 5.8*); e.g., thymic nurse cells may enhance the production of helper T cells from double-positive CD4$^+$,CD8$^+$ thymocytes (16). This lineage oriented differentiation might be related to positive selection by TNC class II molecules (Chapter 4). An individual cell–cell interaction may therefore have more than one functional outcome.

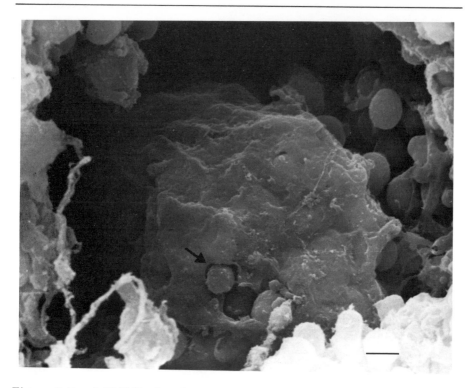

Figure 5.5. A TNC-like lymphoepithelial complex. A lymphoid cell is migrating in or out of this structure (arrow). Scanning electron microscopy, mouse thymus; bar indicates 10 μm. Photograph courtesy of W.van Ewijk, from ref. 6 with permission.

3. Soluble molecules

3.1 Chemotactic factors

Thymocyte precursor cells in the blood are recruited to the thymus by soluble chemotactic molecules (17). Rat subcapsular and perivascular epithelial cells (Type 1 cells) secrete an 11 kDa molecule, that *in vitro,* is specifically chemotactic for the precursors of lymphoid cells. Recently, this molecule has been shown to be β_2 microglobulin, the light chain of MHC class I and CD1 molecules (18). Chemotactic activity has now been observed with rat, mouse, and human β_2 microglobulin, but the mechanism of its function in precursor cell migration is unknown. However, other chemoattractants must also be important since in mice that lack β_2 microglobulin (gene silenced by homologous recombination), migration of cells into the thymus appears to occur normally (19).

3.2 Thymic hormones

Thymic subcapsular and medullary epithelial cells produce a variety of polypeptides that influence progenitor and T cell populations, and have therefore

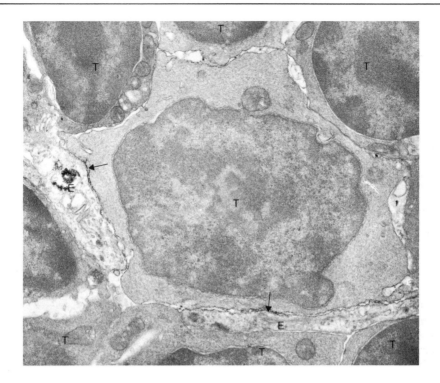

Figure 5.6.   Focal distribution of TCR at point of interaction between thymocyte (T) and cortical epithelial cell (E). Immunoperoxidase labelling of mouse thymus, electron microscopy. Photograph courtesy of A.Farr, from ref. 7 with permission.

been termed 'thymic' hormones. Initially, crude extracts of thymus were shown to have biological activity. Following this, the biologically active fractions were purified by biochemical methods, and in some cases the amino acid sequence of the active component was then determined. Those that have been well defined are described here; many others may exist. Those analysed so far are highly conserved between different species and act across a wide variety of different species barriers (e.g. mouse, rat, man, pig, and calf). However, although the different hormones are often produced by the same cells and appear to have overlapping functions, they show no structural homology with each other and hence cannot be considered as a 'family' of hormones. In general, thymic hormones have been reported to have two main areas of action: phenotypic maturation of progenitor cells from the bone marrow and modulation of the functions of mature T cells. They therefore appear to have both a pre-thymic and a peripheral sphere of influence. Whether these molecules do actually influence intra-thymic T cell development is currently unproven, although their production by subcapsular epithelium suggests that they might act on the prothymocytes that are located in this region of the thymus.

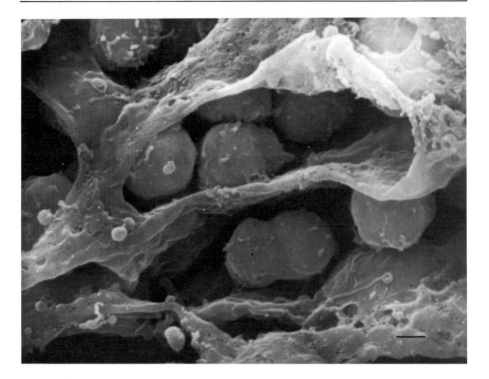

Figure 5.7.   Thymic cortex. Lymphoid cells proliferate in between the processes of epithelial cells. Scanning electron microscopy, mouse thymus; bar represents 2 μm. Photograph courtesy of W.van Ewijk, from ref. 6 with permission.

3.2.1 Molecular characteristics

Despite some uncertainty as to their biological function (Section 3.2.2) the structure of thymic hormones has been studied in considerable detail (20) (*Table 5.1*).

Thymosin fraction 5 was one of the first preparations to be isolated (21). This fraction contains a number of polypeptides ranging in size from 1–15 kDa; of these, thymosin α_1, β_4, and α_7 are the best characterized.

Thymulin (formerly facteur thymique serique, FTS) was first identified in extracts of porcine serum, and later calf thymus. The molecule has been purified, sequenced, and many synthetic analogues produced (20). It is a nonapeptide with a molecular weight of 875 Da, and depends upon the presence of zinc for its biological activity (induction of Thy-1 on murine bone marrow progenitor cells; Section 3.2.2). Although produced by thymic epithelial cells, thymulin is present in the circulation in femtogram quantities (young adult rats and mice, approximately 600 fg/ml plasma; in man <20 years, approximately 1500 fg/ml; in man 20–50 years, 370 fg/ml) (22). It is not detectable in the blood of nude

Figure 5.8.   Thymic cortex. Lymphoid cells differentiate in between the processes of epithelial cells, as shown by expression of microvilli on one thymocyte. Scanning electron microscopy, mouse thymus; bar represents 2 μm. Photograph courtesy of W.van Ewijk, from ref. 6 with permission.

(congenitally athymic) or thymectomized mice and rats (20). Thymulin binds with high affinity (kDa of 10^{-9} and 10^{-7}) to the surface of T cells.

Thymopoietin was initially detected by its inhibitory effect on neuromuscular transmission in studies of myasthenia gravis. It is a 49 amino acid polypeptide, with a molecular weight of 5562 Da. Of its 49 amino acids, a polypeptide (residues 32 – 36; termed thymopentin, TP-5) contains the biological activity (23). TP-5 is identical for both bovine and human thymopoietin.

Thymic humoral factor (THF), molecular weight 3200, was identified by its ability to restore the capacity of spleen cells from neonatally thymectomized donors to induce an *in vivo* graft-versus-host reaction (24).

3.2.2 Biological functions

Many effects have been attributed to thymic hormones (21,25). These fall into two main categories: phenotypic modulation of cells in the progenitor fraction of bone marrow; and functional effects on the peripheral immune system such as enhancement or suppression of peripheral T cell responses, and restoration

Table 5.1.   Thymic hormones

Name	Structure	Reported biological effects
Thymopoietin	MW 5562 49 aa sequenced	↑ Graft rejection ↑ Generation of Tc ↓ Autoimmunity
TP-5	5 aa fragment of thymopoietin	↑ Graft rejection ↑ Generation of Tc ↓ Autoimmunity
Thymosin α_1	MW 3108 28 aa sequenced	↑ Mitogen responses ↑ Antibody production ↑ MIF, γIfn production by T cells ↑ Expression of Thy-1 (mouse) and CD4
Thymosin β_4	MW 4982 43 aa sequenced	↑ TdT expression
Thymosin α_7	MW 2500	↑ Ts *in vitro*
Thymic humoral factor (THF)	MW 3200 sequenced	↑ G-v-HR, MLR, mitogen responses in nude mice
Thymulin	MW 857 (922 with Zn)	Requires Zn for activity Low levels: ↑ immune responses High levels: ↓ immune responses (antibody production, DTH, IL-2 secretion, graft rejection)

MW, molecular weight; aa, amino acids; G-v-HR, graft versus host reaction, measure of T cell immunity; MLR, mixed lymphocyte reaction, measure of T cell immunity.

of T cell function in thymus-deprived (nude or thymectomized) mice. However, few experiments have directly addressed the role of thymic hormones within the thymus itself.

A common feature of all the above polypeptides is their ability to induce phenotypic changes in enriched populations of progenitor cells, isolated from bone marrow by albumin gradient separation (26). Expression of Thy-1, CD5 (Ly-1), and CD8 (Ly-2,3) is rapidly induced on the surface of murine cells, and nuclear TdT is also increased. However, the speed of appearance of these molecules (e.g. 10 min for Thy-1) suggests utilization of a pre-existing intracellular pool rather than *de novo* synthesis. Similar phenotypic changes can be induced by analogues of cyclic AMP, indicating that cAMP-mediated second messenger systems are involved in this aspect of thymic hormone action.

Thymic hormones have been reported to have a wide range of effects on peripheral T cells. *In vitro* assays have demonstrated enhancement of IL-2 secretion (thymulin), PHA and Con A responsiveness (THF, thymosin α_1), and production of macrophage migration inhibition factor, γ-interferon and lymphotoxin/TNF-α (thymosin α_1). Administration of thymulin *in vivo* can enhance the generation of cytotoxic T cells, increase the production of antibody to sheep red blood cells (SRBC; a T dependent antigen) and restore delayed type hypersensitivity (DTH) responses in thymectomized mice. TP-5 enhances both cytotoxic T cell generation and allograft rejection. However, the *in vivo* effects of thymic hormones appear to be dose dependent, since at high doses thymulin has an inhibitory effect on antibody production (SRBC), DTH responses, and allograft rejection. A relative increase in CD8 positive T cells has led to the speculation that circulating levels of thymic hormones may enhance the production of T suppressor cells.

Despite this plethora of peripheral immune effects, there is little evidence of a direct role for thymic hormones within the thymus. However, the fact that high affinity binding sites for thymulin are present on human acute lymphoblastic T cell lines that have an immature thymocyte phenotype (CCRF-HSB2; TdT$^+$, CD5$^+$, CD3$^-$, CD4$^-$,CD8$^-$), strongly suggests that this hormone, at least, may be important in the early stages of intra-thymic T cell development (27). A similar role of thymosin β_4 is indicated by its ability to enhance TdT expression in thymocytes from immunosuppressed mice.

3.2.3 Regulation of secretion

The secretion of thymic hormones is influenced by levels of the thyroid hormones, pituitary hormones, prolactin, and growth hormones (28,29), elevations in these hormones causing increased secretory activity. The reciprocal action has also been documented (30) indicating a possible feedback mechanism. Plasma thymulin levels also exhibit a circadian rhythm, with the highest levels attained in the early morning and the lowest in the late afternoon. The significance of this is unclear at present (Kendall and Safieh, personal observations).

3.3 Cytokines

Cytokines are a group of proteins and glycoproteins that are produced mainly by lymphocytes and cells of the dendritic and monocyte/macrophage lineages, although other cell types are increasingly being recognized as sites of production (*Table 5.2*) (31,32,33). Their major function is to mediate and regulate the interaction both of cells within the immune system as well as between the immune and other organ systems; their action is mediated via binding to specific receptors on the surface of the target cell. Cytokines have been shown to be critical to the functioning of mature T and B lymphocytes in immune responses, where they can act as both growth and maturation factors; they are also of crucial importance in the maturation and differentiation of the B lymphocyte lineage. It is therefore likely that these molecules are also important in T lymphocyte differentiation within the thymus.

In contrast to the thymic hormones where their presence in the serum suggests systemic effects, the action of cytokines is thought to be highly localized, with functionally effective concentrations being found only in the vicinity of the interacting producer and responder cells. This intimacy of interaction is enhanced by the directional secretion of cytokines, whereby they are only released in the

Table 5.2.   Cytokines

Cytokine	Made by	Receptors on*	Functional effects
IL-1	Thymocytes Thymic epithelium T cells B cells Macrophage/DC Endothelium Fibroblasts Keratinocytes	Thymocytes (including DN) T cells Haemopoietic stem cell Macrophages Nerves Endothelium B cells Epidermis	Multiple, including: enhanced secretion of IL-1, IL-2, IL-4, IL-6, TNF-α, GM-CSF, and hormones of hypothalamic, pituitary, and adrenal axis
IL-2	Thymocytes T cells	Thymocytes (DN) Medullary thymocytes T cells B cells Macrophages LAK cells	Proliferation and maturation of T cells, B cells, and macrophages
IL-4	Thymocytes T cells Mast cells Bone marrow stroma Thymic epithelium	Thymocytes (including DN) T cells B cells Thymic epithelium Macrophages/DC Epithelium Fibroblasts Mast cells LAK cells	T and B cell proliferation and maturation e.g. pCTL to CTL Isotype switch to IgE Enhances CD23 Enhances MHC class II Macrophage activation Mast cell activation
IL-6	Thymocytes T cells Macrophages/DC Endothelium Fibroblasts Polymorphs	Thymocytes Thymic epithelium T cells B cells Endothelium	T cell proliferation and differentiation
IL-7	Thymic stroma Bone marrow stroma	Thymocytes (DN) Pro-B cells Pre-B cells	Proliferation of T and B progenitor cells
TNF-α	Macrophages Activated T cells	Thymocytes T cells B cells Endothelial cells Macrophages Polymorphs	Proliferation of T and B cells Enhances MHC class I expression
M-CSF	Thymic epithelium Endothelium Macrophages	Macrophage progenitors	Differentiation of monocyte/macrophage lineage Induces IL-1 secretion
γIFN	T cells	T cells B cells Macrophages Endothelial cells LAK cells	Maturation of CTL Enhances MHC class I and II expression Isotype switch to IgG Activates macrophages

*Receptors for IL-2, IL-4, and IL-7 are also found as soluble molecules which may themselves act as immunoregulatory molecules. DC, dendritic cell.

area of contact between the two cells (34). Thus, although many cytokines are produced by several different cell types and have several different cellular targets, specificity of action is maintained. This specificity is increased by the regulated expression of cytokine receptors. These molecules often increase both in number and affinity after stimulation of the target cell (e.g. IL-2R after antigen-activation of T lymphocytes). Cytokines may also regulate the expression of their own and other cytokine receptors.

Experimental studies have suggested three main areas of intra-thymic T cell development that may be influenced by cytokines: proliferation, phenotypic maturation, and repertoire selection. In addition, since many stromal cells (epithelial, macrophage, dendritic, endothelial, and neuronal) express cytokine receptors, these soluble mediators may also be important in the development, maintenance, and functioning of the thymic microenvironment.

A major function of most cytokines (listed in *Table 5.2*) is to stimulate thymocyte proliferation; however, they differ in their target cell subset and in the co-stimulus required (if any). IL-7 is probably the earliest to act in the thymus. This stromal cell-produced cytokine induces proliferation of early fetal (murine day 14) and adult double-negative thymocytes, without any requirement for a co-stimulus (33). IL-4 may also be important as an autocrine growth factor during early ontogeny since a high proportion of 14 day fetal thymocytes will secrete IL-4 in response to stimulation with phorbol ester and ionomycin, although the physiological ligands that this stimulation mimics are unkown (35).

The role of IL-2 within the thymus is controversial, despite extensive studies on its effects on mature peripheral T lymphocytes. A small proportion of murine thymocytes belonging to the double-negative immature subset express IL-2 receptors (p55 chain, low affinity), although no equivalent immature population has been identified in the rat (36,37). Administration of exogenous IL-2 to mice results in intra-thymic proliferation; and anti-IL-2 receptor antibodies have, in some experiments, blocked fetal thymocyte proliferation (38,39,40). The importance of IL-2 in early thymopoiesis is further suggested by the observation that the IL-2R$^+$ DN thymocytes exhibit greater potential to differentiate to mature T cells than do the IL-2R$^-$/DN population. The source of intra-thymic IL-2 may be the developing thymocytes themselves (autocrine stimulation) or mature antigen-activated T cells that are known to migrate back into the thymus medulla from the periphery. The small number of IL-2R$^+$ medullary thymocytes that have been observed in human and rat thymus may represent this latter, mature population.

IL-6 is co-mitogenic (with PHA) for thymocytes. This cytokine is produced by many cell types including T cells and macrophages and may be responsible for the latter's proliferative effects on developing thymocytes. IL-6 synergizes with both IL-2 and IL-4 to produce enhanced proliferation, and therefore may be important for preparing/priming cells for subsequent developmental signals (e.g. by inducing expression of IL-2R on thymocytes) (41).

The ubiquitous cytokines IL-1 and IL-6 both induce proliferation of thymocytes in the presence of a co-stimulus such as mitogen or combined ionomycin and phorbol ester. The target cell for IL-1 is known to be the double-negative subset (42). The action of the co-stimuli is probably to mimic a physiological interaction of the developing T cell with its microenvironment. Since thymic epithelium is a source of IL-1 and the target cell for IL-1 is the double-negative thymocyte subset, this interaction may involve the subcapsular epithelium – although macrophage-produced IL-1 could act throughout the thymus, and T cell secreted IL-1 can also act as an autocrine growth factor (43,44).

The evidence implicating cytokines in the phenotypic differentiation of thymocytes is more limited. IL-2 induces the development of $CD3^+$, α/β T cell receptor positive T cells, while IL-4 enhances the expression of Thy-1 and high molecular weight CD45 antigen (associated with functionally mature, but naive, thymocytes) on murine thymocytes. In the periphery, IL-4 upregulates CD8 expression, and so might have a similar effect in the thymus (45).

Lymphokines may also be involved in repertoire selection. It has also been proposed that IL-4R on cortical epithelial cells provides a reservoir of surface bound IL-4 that could give a signal for survival to developing thymocytes that interact closely with self-MHC antigens which are also present on the surface of these cells (positive selection) (46,47). In contrast, IL-1 secreted by antigen presenting cells (dendritic cells and macrophages) forms a part of the activation signal that results in the apoptosis of auto-reactive T cells within the thymus (Chapter 4).

Finally, many of these cytokines may be of considerable importance to the cells of the thymic microenvironment. Gamma interferon enhances the expression of MHC class II antigens on thymic epithelial cells (48,49). IL-2 induces proliferation and maturation of thymic macrophages, while IL-1, IL-4, and IL-6 have the potential to act on many microenvironmental cell types such as macrophages, epithelium, endothelium, and nerves (see *Table 5.2* for distribution of cytokine receptors). The interaction between developing T cells and their microenvironment is therefore likely to be bidirectional, such that not only do the lymphocytes receive microenvironmentally-produced signals necessary for their development, but also the cells of the microenvironment receive lymphocyte-derived signals that are required for their normal development, maintenance, and functioning.

It thus seems likely that cytokines play a crucial (but poorly understood) role within the thymus. However, now that the cytokine genes and, more recently, those of some of their receptors have been cloned it should be possible to unravel their individual functions and networks of interaction within the thymus, both at the level of the developing lymphocytes and their microenvironment.

4. Further reading

Austyn,J.M. (1989) *Antigen-presenting Cells* (In Focus series). IRL Press, Oxford.

Hamblin,A.S. (1988) *Lymphokines* (In Focus seies). IRL Press, Oxford.

Kendall,M.D. and Ritter,M.A. (ed.) (1988) *The Microenvironment of the Human Thymus. Thymus Update,* (Annual Review Series), Vol. 1, Harwood Academic, London.

Kyewski,B.A. (1988) Unravelling the complexity of intrathymic cell – cell interactions. *A.P.I.M.S.,* **96**, 1049.

Mosmann,T.R. (1988) Directional release of lymphokines from T cells. *Immunol. Today,* **9**, 306.

5. References

1. Miller,J.F.A.P. (1961) *Lancet,* **ii**, 748.
2. Stutman,O., Yunis,E.J., and Good,R.A. (1969) *Transplantation Proc.,* **i**, 614.
3. von Gaudecker,B. (1990) *Anat. Embryol.,* **183**, 1.
4. Kendall,M.D. (1990) In *The Role of the Thymus in Tolerance Induction. Thymus Update,* (Annual Review Series), Vol. 3, p. 47, Harwood Academic, London.
5. Wekerle,H., Ketelsen,U.P., and Ernst,M. (1980) *J. Exp. Med.,* **151**, 925.
6. van Ewijk,W. (1988) *Lab. Invest.,* **59**, 579.
7. Farr,A.G., Anderson,S.K., Marrack,P., and Kappler,J. (1985) *Cell,* **43**, 543.
8. Lo,D. and Sprent,J. (1986) *Nature,* **319**, 672.
9. Benoist,C. and Mathis,D. (1989) *Cell,* **58**, 1027.
10. van Ewijk,W., Ron,Y., Manaco,J., Kappler,J., Marrack,P., Le Meur,M., Gerlinger,P., Durand,B., Benoist,C., and Mathis,D. (1988) *Cell,* **53**, 357.
11. Ready,A.R., Jenkinson,E.J., Kingston,R., and Owen,J.J.T. (1984) *Nature,* **310**, 231.
12. Nabarra,B. and Papiernik,M. (1988) *Lab. Invest.,* **58**, 524.
13. Kyewski,B., Momburg,F., and Schirrmacker,V. (1987) *Eur. J. Immunol.,* **17**, 961.
14. Kyewski,B.A. (1988) *A.P.M.I.S.,* **96**, 1049.
15. Denning,S.M., Tuck,D.T., Vollger,L.W., Springer,K.H., and Haynes,B.F. (1987) *J. Immunol.,* **139**, 2573.
16. Andrews,P., Boyd,R.L., and Shortman,K. (1985) *Eur. J. Immunol.,* **15**, 1043.
17. Imhof,B.A., Dengnier,M.A., Girault,J.M., Champion,S., Damais,C., Itoh,T., and Thiery,J.P. (1988) *Proc. Natl Acad. Sci.,* **85**, 7699.
18. Dargemont,C., Dunon,D., Deugnier,M., Denoyelle,M., Girault,J., Lederer,F., Ho Diep Le,K., Godean,F., Thiery,J., and Imhof,B.A. (1989) *Science,* **246**, 803.
19. Zijlstra,M., Bix,M., Simister,N.E., Loring,J.M., Raulet,D.H., and Jaenisch,R. (1990) *Nature,* **344**, 742.
20. Dardenne,M. (1988) In *The Human Thymic Microenvironment, Thymus Update,* (Annual Review Series), Vol. 1, p. 101, Harwood Academic, London.
21. Low.,T.L.K., Thurman,G.B., McAdoo,M., McClure,J., Rossio,L., Naylor,P.H., and Goldstein,A.L. (1979) *J. Biol. Chem.,* **254**, 981.
22. Safieh,B., Kendall,M.D., Norman,J.C., Metreau,E., Dardenne,M., Bach,J.F., and Pleau,J.M. (1990) *J. Immunol. Meth.,* **127**, 255.
23. Goldstein,G., Scheid,M.P., Boyse,E.A., Schlesinger,D.H., and van Vauwe,J. (1979) *Science,* **204**, 1309.
24. Trainin,N., Rotter,V., Yakir,Y., Leve,R., Handzel,Z., Shahat,B., and Zaizov,R. (1979) *Annal. N.Y. Acad. Sci.,* **332**, 9.
25. Goldstein,A.L., Low,T.L.K., Thurman,G.B., Zatz,M.M., Hall,N., Chen,J., Hu,S.-K., Naylor,P.B., and McClure,J.E. *Rec. Prog. Horm. Res.,* **37**, 369.
26. Komuro,K. and Boyse,E.A. (1973) *J. Exp. Med.,* **138**, 479.
27. Pleau,J.-M., Fuentes,V., Morgat,J.-L., and Bach,J.-F. (1980) *Proc. Natl Acad. Sci. USA,* **77**, 2861.
28. Dardene,M., Savino,W., Gagnerault,M.-C., Itoh,T., and Bach,J.-F. (1989) *Endocrinology,* **125**, 3.
29. Dardene,M. and Savino,W. (1990) *Prog. NeuroEndocrinol.,* **3**, 18.
30. Spangelo,B.L., Judd,A.M., Ross,P.C., Login,I.S., Jarvis,W.D., Badamchian,M., Goldstein,A.L., and MacLeod,R.I. (1987) *J. Endocrinol.,* **121**, 2035.
31. Galvani,D.W. (1988) *J. Roy. Coll. Phys.,* **22**, 226.
32. Balkwill,F.R. and Burke,F. (1989) *Immunol. Today,* **10**, 299.
33. Henney,C.S. (1989) *Immunol. Today,* **10**, 170.
34. Poo,W.-J., Conrad,L., and Janeway Jr.,C.A. (1988) *Nature,* **332**, 378.
35. Sideras,P., Funa,K., Zalcbere-Quintana,I., Xanthopoulos,K.G., Kisielow,P., and Palacios,R. (1988) *Proc. Natl Acad. Sci. USA,* **85**, 218.
36. Shimonkevitz,R.P., Husmann,L.A., Bevan,M.J., and Crispe,I.N. (1987) *Nature,* **329**, 157.

37. Takacs,L., Ruscetti,F.W., Kovacas,E.J., Rocha,B., Brocke,S., Diamantstein,T., and Mathieson,B.J. (1988) *J. Immunol.,* **141**, 3810.

38. Jenkinson,E.J., Kingston,R., and Owen,J.J.T. (1987) *Nature,* **329**, 160.

39. Plum,J. and De Smedt,M. (1988) *Eur. J. Immunol.,* **18**, 795.

40. Gearing,A.J.H., Wadhwa,M., and Perris,A.D. (1986) *Eur. J. Immunol.,* **16**, 1171.

41. Hodgkin,P.D., Bond,M.W., O'Garra,A., Frank,G., Lee,F., Coffman,R.L., Zlotnik,A., and Howard,M. (1988) *J. Immunol.,* **141**, 151.

42. Chaudhri,G., Clark,I.A., and Ceredig,R. (1988) *Clin. Exp. Immunol.,* **73**, 51.

43. Le,P.T., Tuck,D.T., Dinarello,C.A., Haynes,B.F., and Singer,K.H. (1987) *J. Immunol.,* **138**, 2520.

44. Tartakovsky,B., Finnegan,A., Muegge,K., Brody,D.T., Kovacs,E.J., Smith,M.R., Berzofsky,J.A., Young,H.A., and Durum,S.K. (1988) *J. Immunol.,* **41**, 3863.

45. Paliard,X., de Waal Malefijt,R., de Vries,J.E., and Spits,H. (1988) *Nature,* **335**, 642.

46. Larche,M., Lamb,J.R., O'Hehir,R.E., Imami-Shita,N., Zanders,E.D., Quint,D.E., Moqbel,R., and Ritter,M.A. (1988) *Immunology,* **65**, 617.

47. von Gaudecker,B., Larche,M., Schuurman,H., and Ritter,M.A. (1989) *Thymus,* **13**, 187.

48. Berrih,S., Arenzana-Seisdedos,F., Cohen,S., Devos,R., Charron,D., and Virelizier,J. (1985) *J. Immunol.,* **135**, 1165.

49. Rocha,B., Lehuen,A., and Papiernik,M. (1988) *J. Immunol.,* **140**, 1076.

6

Unresolved issues in thymus research

In the previous chapters, we have described the structure and function of the thymus as it is currently understood. It is now clear that the thymus is seeded by small numbers of progenitor cells, which undergo multiple rounds of cell division, rearrange the genes for their T cell antigen receptors and express these receptors at the cell surface. They are then subjected to a selection process, which permits the survival of T cells which are self-tolerant and self-MHC restricted. These cells acquire functional competence, and exit to the periphery as primary T cells (*Figure 6.1*). However, the thymus remains an area of active research, and many key questions remain unanswered. This chapter is an attempt to spell out the main issues with which thymus research is now preoccupied.

1. The commitment of progenitor cells

Thymocytes of both α/β and γ/δ lineages are derived from haemopoietic stem cells. These stem cells have been highly purified from adult bone marrow, and extensively characterized; but their developmental potential may not be the same at all stages of life. Two populations of lymphocytes, the CD5$^+$ B cells and dendritic epidermal γ/δ cells, appear early in life and cannot be replaced by reconstitution with adult bone marrow. This may be because of differences in the fetal and adult environments; or it may be because of irreversible maturation of the stem cell pool. In this case, extrapolation from the development of the fetal thymus to the steady state cell turnover in the adult would be invalid.

It is not clear whether the cells which arrive in the fetal or the adult thymus are committed to the T cell lineage, or are multipotential cells which retain the ability to give rise to other haemopoietic populations. There is evidence for the expression of T cell specific genes and cell surface molecules in the fetal liver, and in the tissue surrounding the developing thymus, but the relationship of these events to irreversible lineage commitment is unknown. It remains possible that the thymus is seeded by multilineage haemopoietic stem cells; or by partly

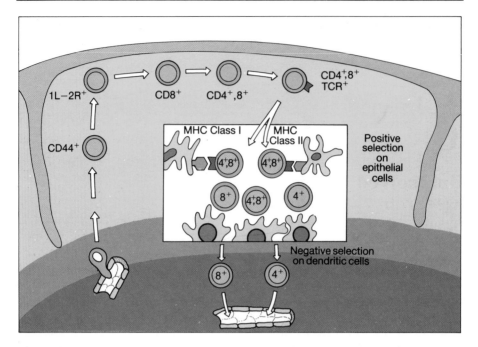

Figure 6.1.   Schematic summary of T cell development: maturation, selection, and migration within the thymus. The chronological sequence of selection events is unknown and subject to debate. IL-2R p55 expression is best defined in the mouse.

committed pro-lymphocytes; or by committed pro-T cells. Even the lineage split between α/β and γ/δ lineage cells has not been firmly placed in the thymus.

2. The control of TCR gene rearrangement

An early stage of maturation involves rearrangement of the γ locus, and the δ segment of the α/δ locus, although these steps may be abortive or bypassed altogether in cells which are destined for the α/β T cell lineage. The regulation of the sequence of TCR gene rearrangement is poorly understood, but one current idea is that the DNA flanking the TCR genes contains silencer as well as enhancer elements, and that these may transduce signals to abort rearrangement of a gene segment, such as the δ region in committed α/β T-lineage cells.

   A deeper problem is the control of the selective rearrangement of different TCR genes as development proceeds. Are δ genes embedded in the α locus in all species, and does this alone account for the sequential rearrangement and expression of first γ/δ, and then α/β receptors? In TCR γ genes, as in IgH genes, the pattern of V gene rearrangement and expression favours some V genes in the early fetus, others later in fetal life, and still others in adult animals. Is this an epiphenomenon of the mechanics of gene rearrangement, or a response to the changing priorities of the immune system as development proceeds? If the latter, how is recombination directed to act on a specified V gene?

3. The role of lymphokines

Cells of the α/β lineage transiently express the p55 chain of the IL-2R, which then disappears from the cell surface before the expression of CD4 or CD8. Mysteriously this subset of cells will not respond to IL-2 *in vitro*, and although their receptors will bind to IL-2 they are not then internalized. So why is the IL-2R p55 expressed? Functional experiments in fetal thymus organ culture, and *in vivo*, suggest that antibodies against p55 can inhibit T cell differentiation, but strangely there seems to be abrogation of the final maturation step to CD4$^+$ or CD8$^+$ single-positive cells. This only makes sense if the IL-2R p55 transmits a signal which has consequences later, perhaps at the time of positive selection. One obvious target is the rearrangement of the α regions of the α/δ locus, but this has not yet been examined. The β locus is probably fully rearranged by this stage, since cells with surface IL-2R p55 express full-length β chain message; but the β chains are not expressed at the cell surface until the productive rearrangement of the α locus is complete.

Both IL-2 and IL-4 mRNA are expressed in a developmentally regulated fashion in the early fetal thymus, in a 'spike' around fetal day 13 – 15, just when γ/δ cells are most prominent. The prevalence of IL-4 receptors at low density on essentially all human thymocytes argues for the importance of this lymphokine in α/β T cell development; but the role of IL-4 remains unknown.

4. The function of thymic stromal cells

The non-lymphoid component of the thymus consists of at least five types of epithelial cells, plus macrophages, dendritic cells and a very small number of B cells. The experimental procedures currently available permit the discrimination of effects due to radiation-sensitive versus radiation-resistant cells, bone marrow-derived versus other cells, deoxyguanosine-resistant epithelium, and the cortical versus the medullary epithelium as defined by variants of the H-2E α gene promoter. So far these techniques have provided a broadly self-consistent picture of the roles of cortical epithelial cells versus medullary epithelia and dendritic cells in thymocyte selection.

One key problem is to discriminate the maturation events controlled by various types of cortical epithelial cells. Which of these cells support or participate in early thymocyte proliferation, TCR gene rearrangement, and the expression of CD4 and CD8? And is this participation mediated by the production of growth factors, or by specific cell – cell interactions? Evidence so far is limited to provocative observations of adhesion molecules, lymphokines, and lymphokine receptors on stromal cells and stroma-derived cell lines. The future study of this problem appears likely to be intimately intertwined with efforts to reconstitute thymic microenvironments *in vitro*.

5. When does positive selection occur?

Positive selection has two consequences; cells with self-MHC restriction potential preferentially survive, and cells with MHC class I specificity develop into the CD8$^+$ mature subset, while class II restricted cells become CD4$^+$. These two effects could be aspects of a single cell interaction, or they could be separate. One 'instructive' model proposes that positive selection acts on CD4$^+$,CD8$^+$ cortical cells. In this model, recognition of self-MHC class I on a cortical epithelial cell results in both a message which causes cell survival, and selective down-regulation of CD4 expression to give a mature CD8$^+$ cell; conversely, recognition of self-MHC class II leads to cell survival and down-regulation of CD8, to yield a functional CD4$^+$ cell. This implies distinct intracellular signalling mechanisms in class I versus class II recognition.

An alternative, 'selective' model is that T cells become either CD4$^+$ or CD8$^+$ at random, and then receive a survival signal at this stage if their restriction specificity matches the accessory molecule expressed. This model is more economical in terms of molecular processes, since it depends only on the transmission of a survival signal, and there is no need for different intracellular signals during class I versus class II recognition. However, the model implies that class II restricted T cells should have a short life in the CD8$^+$ single-positive subset before they die, and vice versa for class I restricted CD4$^+$ cells. In fact the cell death observed in the thymus seems to affect CD4$^+$,CD8$^+$ double-positive cells, and in transgenic mice with a uniform TCR of known restriction specificity the moribund cells with the wrong CD4/CD8 phenotype are not seen.

6. The TCR ligand during positive selection

Both models of positive selection have to deal with the nature of the ligand recognized during this process. MHC-restricted recognition of non-self antigens by T cells in the periphery involves presentation of the antigen as peptides, embedded in a groove between two α-helical regions on the membrane-distal face of an MHC molecule. The apparent paradox is that T cells which respond to self-MHC plus self-peptides are subject to clonal deletion, yet positive selection favours the development of cells which will respond to self-MHC plus non-self peptides in the periphery.

There are two general categories of explanation for this effect. One is that positive selection interactions work at very low affinity, so that T cells with a very weak, but non-zero affinity for self-peptides plus self-MHC are selected. In this model, the affinity thresholds for clonal deletion and for T cell activation in the periphery are much higher than the threshold for positive selection in the thymus, so that strongly self-reactive cells will be clipped out of the repertoire by clonal deletion but weakly self-reactive cells will escape deletion, and will not be activated in the periphery. The theory then proposes that weakly self-reactive TCR will cross-react strongly with self-MHC plus non-self peptides. In the light of the structural information now available on MHC class I molecules, this model has been restyled by proposing either that thymus epithelial cells express 'naked MHC', with an empty antigen-binding groove, or that an inert 'selecting peptide' is present, but does not interact with the TCR during positive selection. In either case, the low affinity results from interactions only on the flanking α-helices, and not across the whole face of the peptide-MHC complex.

The other general class of models of positive selection proposes that the process involves high-affinity interactions between the TCR and a set of peptides unique to the positively-selecting thymus epithelial cells. This model is attractive because it resolves the double handicap of low affinity and low TCR density during positive selection. In this model, positively selected cells are not deleted because they were selected on peptide-MHC complexes different from those encountered on the intrathymic cells which cause clonal deletion, and likewise different from 'self' encountered in the periphery. Cells selected on thymic epithelial self-peptides plus self-MHC are then proposed to cross-react on antigenic non-self peptides plus self-MHC. In this model, the thymic epithelial peptides form an internal image which schools immature T cells for the world of external antigens.

7. The timing of clonal deletion

The positive and negative selection of developing T cells depends on ligation of the TCR. So how does a T cell know if it is being positively or negatively selected? There are two general ideas. The process could reflect the T cell progenitor maturation programme, so that susceptibility to positive selection and deletion occur at different maturational stages. In support of this idea, a population of cortical thymocytes exists which is refractory to deletion by ligation of the TCR β chain, but responds to ligation of CD3. This is evidence for two functionally distinct subsets of TCR in the CD4$^+$,CD8$^+$ thymocytes, but not for directly sequential stages with different selectability properties.

Alternatively, developing T cells may respond to quite distinct signals delivered by the thymus epithelium during positive selection, versus bone marrow-derived stromal cells during negative selection. In this case the two kinds of selection need not occur in any particular order. Studies in mice with transgenic TCRs cross-reactive for several specificities have shown clonal deletion of the same T cell receptor at different maturational stages, depending on the antigen. This makes it less likely that positive and negative selection are separated by developmental timing, since clonal deletion (at least) can occur at many stages of T cell maturation.

8. How and where do thymocytes die?

There are at least three reasons why developing T cells might die in the thymus: failure to rearrange a functional set of TCR genes, lack of self-MHC restriction potential, and deletion due to self-tolerance. Since apoptosis is the hallmark of programmed or physiological cell death, it may be that all cell death in the thymus is by this mechanism. However, in thymus sections it is difficult to discern numerous apoptotic cells, which has provoked the suggestion that moribund thymocytes leave the thymus and die unobtrusively, somewhere else. Some evidence on the reutilization of nucleotide analogues after *in vivo* pulse-labelling supports this idea; but some apoptotic nuclei are present in the thymus and in cell suspensions, so the graveyard of moribund thymocytes is not yet settled.

9. Exit from the thymus: a one-way trip?

Newly exported T cells transiently express on their membrane a very low level of the thymocyte antigen HSA, which is down-regulated during functional maturation but not finally lost until around 24 hours after cells reach the periphery. Intrathymic TCR-high CD4$^+$ and CD8$^+$ single-positive cells exist in two populations, HSA-dull cells which are Qa-2 negative, and HSA-negative cells which express some Qa-2 and thus more closely resemble peripheral T cells. It is plausible that only the HSA-dull, Qa-2 negative cells are the true late thymocytes, while the other population are recirculated T cells which have been through the periphery. Since thymocytes in the medulla are still susceptible to clonal deletion, this provides a route whereby returning peripheral T cells could influence the emerging T cell repertoire.

   The above areas are all the subject of much active current research. Thus the answers to many of these important questions may emerge during the next few years, leading to a more complete understanding of the development of T lymphocytes and the true function of the thymus.

Glossary

Adoptive transfer: experimental transfer of functional lymphocytes from one animal to another.

Allogeneic: from genetically distinct members of the same species (i.e. expressing allelic differences), cf. syngeneic.

Anergy: unresponsiveness of antigen-specific T cells after exposure to antigen.

APC: antigen presenting cell. In the context of the thymus, 'professional' APC, such as dendritic cells, are distinguished from 'non-professional' APC, such as T cells, B cells or thymic epithelium, which may present antigen under some circumstances.

Apoptosis: programmed cell death, involving endonuclease fragmentation of DNA. The proposed mechanism of clonal deletion in the thymus.

Blast cell: enlarged cell, early stage of cell division.

Bursectomy: removal of the bursa of Fabricius.

Burst size: the number of progeny derived from one progenitor cell.

CD: cluster of differentiation. WHO approved system of classification of monoclonal anti-leucocyte antibodies and the antigens to which they bind.

CGRP: calcitonin gene related peptide.

Chimera: an animal which contains cells from genetically distinct origins. Radiation chimeras are widely used to study the interactions between radiosensitive haemopoietic cells, and radioresistant tissues of the host.

Clonal deletion: elimination of self-reactive cells.

Clonal selection: determination of the fate of lymphocytes, based on the specificity of their clonally distributed antigen receptors.

Congenic: two related inbred strains, differing only at a defined genetic locus.

Cortex: the main outer region of the thymus, containing immature and non-functional cells.

CTES: cluster of thymic epithelial staining. Preliminary classification of monoclonal anti-epithelial antibodies according to their immunohistochemical staining patterns on thymus sections.

Cytokine: soluble molecules secreted by haemopoietic (and other) cells, which influence the growth and maturation of haemopoietic (and other) cells. 'Hormones which immunologists discovered first'; Gearing (1989) *Thymus Update* **2**, 127.

Dendritic cells: MHC class II positive, bone marrow-derived, antigen-presenting cells with a dendritic morphology, which are abundant in the thymic medulla.

DN: 'double negative' in thymologists' slang means $CD4^-, CD8^-$, unless otherwise qualified.

DP: 'double positive' in thymologists' slang means $CD4^+, CD8^+$, unless otherwise qualified.

Endonuclease: enzyme that catalyses the cleavage of DNA into small fragments (see also apoptosis).

Epitope: antigenic determinant. Smallest unit of an antigen that is recognized by a T cell receptor (or immunoglobulin) binding site.

Flow cytometry: cells in suspension flow through a laser beam, and signals emitted are collected and analysed by computer. Cells can be analysed for physical properties (size, granularity) and for fluorescent labels that identify subpopulations.

Hassall's corpuscle: concentric whorl of epithelial cells located in thymic medulla.

Haemopoietic: refers to the development of all lineages of blood forming cells (lymphocytes, erythroid cells, monocyte/macrophages, neutrophils, basophils/mast cells, megakaryocytes/platelets, dendritic cells).

HEV: high endothelial venule. Venule in which the lining endothelial cells are fat (not flat as in other blood vessels). Site of lymphocyte emigration from blood to tissues, typical of lymphoid organs and inflammatory sites (see also PCV).

Homing: the migration and localization of cells to a specific anatomical site.

Homing receptors: cell surface molecules which recognize ligands at particular anatomical locations, and direct recirculating cells to these sites.

Interdigitating cell: see dendritic cell.

Interleukin: a growth and/or maturation factor produced by and acting on haemopoietic (and other) cells. Currently 10 are recognized.

Medulla: the central anatomical zone of the thymus, containing mainly functionally competent T lymphocytes.

MHC: major histocompatibility complex. The gene complex that encodes the molecules recognized by T cells (MHC class I and II), plus other genes including those for complement components (class III) and tumour necrosis factor (TNF). Allogeneic MHC is highly immunogenic.

MHC restriction: ability of a T cell to see antigen *only* when it is associated with self allelic variants of MHC molecules.

Negative selection: removal of self reactive specificities by cell deletion and other mechanisms (anergy, veto, suppression).

Nurse cell: thymic nurse cell (TNC). An epithelial cell containing an aggregate of thymocytes, isolated from thymus. Similar structures are visible *in vivo* using immunohistochemistry and scanning EM. They may be sites of lymphocyte-stromal cell interactions.

Organ culture: *in vitro* maintenance of an anatomically intact organ. In the present context, fetal thymus lobes are maintained thus and may be used to support precursor cell differentiation.

PCV: post capillary venule (not a helpful term as all venules are by definition 'post capillary'). Venule in which the lining endothelial cells are fat (not flat as in other blood vessels). Site of lymphocyte emigration from blood to tissues, typical of lymphoid organs and inflammatory sites (see also HEV).

Positive selection: shaping of the T cell repertoire to favour the recognition of exogenous antigens in association with self allelic variants of MHC class I or class II molecules.

Precursor cell: one cell which gives rise to another. There are no implications about the lineage diversity or number of its differentiation products.

Progenitor cell: a cell which is committed to one or more lineages and which gives rise to many progeny, but is not self renewing (cf. stem cell).

Proliferation: cell division by mitosis.

Prothymocyte: thymocyte progenitor, located mainly in subcapsular zone.

Repertoire: the diverse antigen receptor recognition specificities of a set of T (or B) cells.

Stem cell: a self-renewing cell which gives rise to more differentiated cell populations. The pluripotent haemopoietic stem cell can repopulate all lineages of haemopoietic cells, and its progeny are competent to survive for the life of the animal.

Subcapsule: narrow outermost zone of thymus that lies just under the connective tissue capsule, containing mainly thymocyte precursor cells.

Suppression: antigen-specific down-regulation of an immune response, often mediated by CD8$^+$ T cells.

Syngeneic: from a genetically identical member of the same species (cf. allogeneic).

Thymectomy: removal of the thymus.

Thymic epithelium: the major non-lymphoid population in the thymus. These cells are rich in cytokeratins, form an interconnecting network surrounding developing thymocytes, and are essential for T cell maturation.

Thymic hormone: a small peptide secreted by thymic epithelium. The term does not imply a known thymic function.

Thymic stroma: all non-T cells of the thymus.

Thymocyte: any T-lineage cell in the thymus.

Thymotaxin: the chemotactic activity of β_2 microglobulin for bone marrow derived cells.

Tolerance: immunological unresponsiveness to self-molecules, or in experimental models of self.

TdT: terminal deoxynucleotidyl transferase. Enzyme that can catalyse the addition of mononucleotides to any 3'-OH terminated segment of DNA in the absence of a complementary template.

Transgenic: an animal which has incorporated foreign DNA into the germline (and hence all cells of the body) as a result of the experimental introduction of DNA into the fertilized ovum.

Veto cell: an antigen-presenting cell which inactivates T cells that recognize antigens on the veto cell surface.

Index